AN INTRODUCTION TO MATRICES, SETS AND GROUPS FOR SCIENCE STUDENTS

AN INTRODUCTION TO

Matrices, Sets and Groups for Science Students

G. STEPHENSON

B.SC., PH.D., D.I.C.

Emeritus Reader in Mathematics
Imperial College of Science and Technology
University of London

Dover Publications, Inc.
New York

Published in Canada by General Publishing Company, Ltd., 30 Lesmill Road, Don Mills, Toronto, Ontario.
Published in the United Kingdom by Constable and Company, Ltd., 10 Orange Street, London WC2H 7EG.

This Dover edition, first published in 1986, is an unabridged and slightly corrected republication of the revised fourth impression (1974) of the work first published by the Longman Group Ltd., London, in 1965.

Manufactured in the United States of America
Dover Publications, Inc., 31 East 2nd Street, Mineola, N.Y. 11501

Library of Congress Cataloging-in-Publication Data

Stephenson, G. (Geoffrey), 1927–
An introduction to matrices, sets and groups for science students.

Reprint. Originally published: London: Longman Group, 1974.
Includes index.
1. Matrices. 2. Set theory. 3. Groups, Theory of. I. Title.
QA188.S69 1986 512.9′434 85-29370
ISBN 0-486-65077-4

TO

JOHN AND LYNN

CONTENTS

5. Eigenvalues and Eigenvectors

6. Diagonalisation of Matrices

7. Functions of Matrices

8. Group Theory

PREFACE

THIS book is written primarily for undergraduate students of science and engineering, and presents an elementary introduction to some of the major branches of modern algebra – namely, matrices, sets and groups. Of these three topics, matrices are of especial importance at undergraduate level, and consequently more space is devoted to their study than to the other two. Nevertheless the subjects are interrelated, and it is hoped that this book will give the student an insight into some of the basic connections between various mathematical concepts as well as teaching him how to manipulate the mathematics itself.

Although matrices and groups, for example, are usually taught to students in their second and third year ancillary mathematics courses, there is no inherent difficulty in the presentation of these subjects which make them intractable in the first year. In the author's opinion more should be done to bring out the importance of algebraic structures early on in an undergraduate course, even if this is at the expense of some of the more routine parts of the differential calculus. Accordingly this book has been made virtually self-contained and relies only on a minimum of mathematical knowledge such as is required for university entrance. It should therefore be suitable for physicists, chemists and engineers at any stage of their degree course.

Various worked examples are given in the text, and problems for the reader to work are included at the end of each chapter. Answers to these problems are at the end of the book. In addition, a list of further reading matter is given which should enable the student to follow the subjects discussed here considerably farther.

The author wishes to express his thanks to Dr. I. N. Baker and Mr. D. Dunn, both of whom have read the manuscript and made numerous criticisms and suggestions which have substantially improved the text. Thanks are also due to Dr. A. N. Gordon for reading the proofs and making his usual comments.

Imperial College, London. G. S.
1964

AN INTRODUCTION TO MATRICES, SETS AND GROUPS FOR SCIENCE STUDENTS

CHAPTER 1

Sets, Mappings and Transformations

1.1 Introduction

The concept of a set of objects is one of the most fundamental in mathematics, and set theory along with mathematical logic may properly be said to lie at the very foundations of mathematics. Although it is not the purpose of this book to delve into the fundamental structure of mathematics, the idea of a set (corresponding as it does with our intuitive notion of a collection) is worth pursuing as it leads naturally on the one hand into such concepts as mappings and transformations from which the matrix idea follows and, on the other, into group theory with its ever growing applications in the physical sciences. Furthermore, sets and mathematical logic are now basic to much of the design of computers and electrical circuits, as well as to the axiomatic formulation of probability theory. In this chapter we develop first just sufficient of elementary set theory and its notation to enable the ideas of mappings and transformations (linear, in particular) to be understood. Linear transformations are then used as a means of introducing matrices, the more formal approach to matrix algebra and matrix calculus being dealt with in the following chapters.

In the later sections of this chapter we again return to set theory, giving a brief account of set algebra together with a few examples of the types of problems in which sets are of use. However, these ideas will not be developed very far; the reader who is interested in the more advanced aspects and applications of set theory should consult some of the texts given in the list of further reading matter at the end of the book.

1.2 Sets

We must first specify what we mean by a set of elements. Any collection of objects, quantities or operators forms a set, each individual object, quantity or operator being called an element (or member) of the set. For example, we might consider a set of students, the set of all real numbers between 0 and 1, the set of electrons in an atom, or the set of operators $\partial/\partial x_1$, $\partial/\partial x_2$, . . ., $\partial/\partial x_n$. If the set contains a finite number of elements it is said to be a finite set, otherwise it is called infinite (e.g. the set of all positive integers).

Sets will be denoted by capital letters A, B, C, . . ., whilst the elements of a set will be denoted by small letters a, b, . . . x, y, z, and sometimes by numbers 1, 2, 3, ...

A set which does not contain any elements is called the empty set (or null set) and is denoted by $\emptyset$. For example, the set of all integers x in $0<x<1$ is an empty set, since there is no integer satisfying this condition. (We remark here that if sets are defined as containing elements then $\emptyset$ can hardly be called a set without introducing an inconsistency. This is not a serious difficulty from our point of view, but illustrates the care needed in forming a definition of such a basic thing as a set.)

The symbol $\in$ is used to denote membership of – or belonging to – a set. For example, $x \in A$ is read as 'the element x belongs to the set A'. Similarly $x \notin A$ is read as 'x does not belong to A' or 'x is not an element of A'.

If we specify a set by enumerating its elements it is usual to enclose the elements in brackets. Thus

$$A = \{2, 4, 6, 8, 10\} \tag{1}$$

is the set of five elements – the numbers 2, 4, 6, 8 and 10. The order of the elements in the brackets is quite irrelevant and we might just as well have written $A = \{4, 8, 6, 2, 10\}$. However, in many cases where the number of elements is large (or not finite) this method of specifying a set is no longer convenient. To overcome this we can specify a set by giving a 'defining property' E (say) so that A is the set of all elements with property E, where E is a well-defined property possessed by some objects. This is written in symbolic form as

$$A = \{x;\ x \text{ has the property } E\}. \tag{2}$$

For example, if A is the set of all odd integers we may write

$$A = \{x;\ x \text{ is an odd integer}\}.$$

This is clearly an infinite set. Likewise,

$$B = \{x;\ x \text{ is a letter of the alphabet}\}$$

is a finite set of twenty-six elements – namely, the letters $a, b, c \ldots y, z$.

Using this notation the null set (or empty set) may be defined as

$$\emptyset = \{x;\ x \neq x\}. \tag{3}$$

We now come to the idea of a subset. If every element of a set A is also an element of a set B, then A is called a subset of B. This is denoted symbolically by $A \subseteq B$, which is read as 'A is contained in B' or 'A is included in B'. The same statement may be written as $B \supseteq A$, which is read as 'B contains A'. For example, if

$$A = \{x;\ x \text{ is an integer}\}$$

and

$$B = \{y;\ y \text{ is a real number}\}$$

then $A \subseteq B$ and $B \supseteq A$. Two sets are said to be equal (or identical) if and only if they have the same elements; we denote equality in the usual way by the equality sign $=$.

We now prove two basic theorems.

Theorem 1. If $A \subseteq B$ and $B \subseteq C$, then $A \subseteq C$.

For suppose that x is an element of A. Then $x \in A$. But $x \in B$ since $A \subseteq B$. Consequently $x \in C$ since $B \subseteq C$. Hence every element of A is contained in C – that is, $A \subseteq C$.

Theorem 2. If $A \subseteq B$ and $B \subseteq A$, then $A = B$.

Let $x \in A$ (x is a member of A). Then $x \in B$ since $A \subseteq B$. But if $x \in B$ then $x \in A$ since $B \subseteq A$. Hence A and B have the same elements and consequently are identical sets – that is, $A = B$.

If a set A is a subset of B and at least one element of B is not an element of A, then A is called a proper subset of B. We denote this by $A \subset B$. For example, if B is the set of numbers $\{1, 2, 3\}$ then the sets $\{1, 2\}$, $\{2, 3\}$, $\{3, 1\}$, $\{1\}$, $\{2\}$, $\{3\}$ are proper subsets of B. The empty set $\emptyset$ is also counted as a proper subset of B, whilst the set $\{1, 2, 3\}$ is a subset of itself but is not a proper subset. Counting proper subsets and subsets together we see that B has eight subsets. We can now show that a set of n elements has 2^n subsets. To do this we simply sum the number of ways of taking r elements at a time from n elements. This is equal to

$$\sum_{r=0}^{n} {}^{n}C_r = {}^{n}C_0 + {}^{n}C_1 + \ldots + {}^{n}C_n = 2^n \tag{4}$$

using the binomial theorem. This number includes the null set (the ${}^{n}C_0$ term) and the set itself (the ${}^{n}C_n$ term).

1.3 Venn diagrams

A simple device of help in set theory is the Venn diagram. Fuller use will be made of these diagrams in 1.7 when set operations are considered in more detail. However, it is convenient to introduce the essential features of Venn diagrams at this point as they will be used in the next section to illustrate the idea of a mapping.

The Venn diagram method represents a set by a simple plane area, usually bounded by a circle – although the shape of the boundary is quite irrelevant. The elements of the set are represented by points inside the circle. For example, suppose A is a proper subset of B (i.e. $A \subset B$). Then this can be denoted by any of the diagrams of Fig. 1.1.

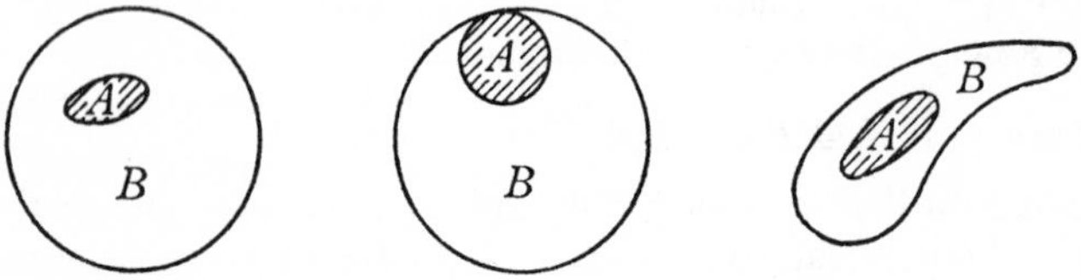

Fig. 1.1

If A and B are sets with no elements in common – that is no element of A is in B and no element of B is in A – then the sets are said to be disjoint. For example, if

$$A = \{x;\ x \text{ is a planet}\}$$

and

$$B = \{y;\ y \text{ is a star}\}$$

then A and B are disjoint sets. The Venn diagram appropriate to this case is made up of two bounded regions with no points in common (see Fig. 1.2).

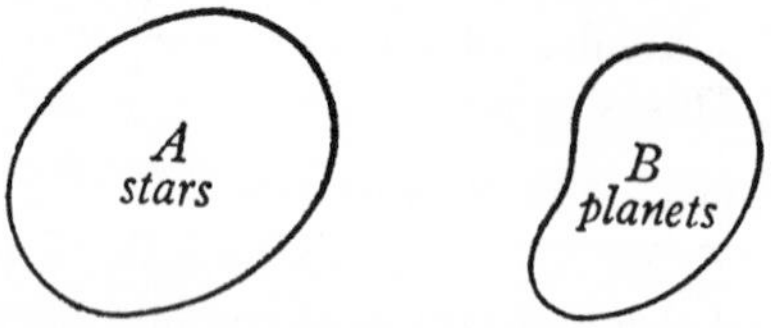

Fig. 1.2

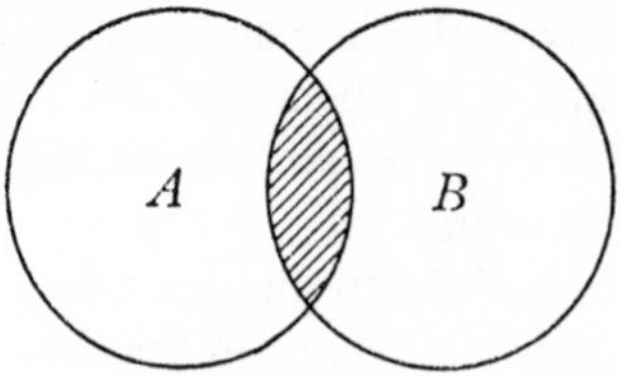

Fig. 1.3

It is also possible to have two sets with some elements in common. This is represented in Venn diagram form by Fig. 1.3, where the shaded region is common to both sets. More will be said about this case in 1.7.

1.4 Mappings

One of the basic ideas in mathematics is that of a mapping. A mapping of a set A onto a set B is defined by a rule or operation which assigns to every element of A a definte element of B (we shall see later that A and B need not necessarily be different sets). It is commonplace to refer to mappings also as transformations or functions, and to denote a mapping f of A onto B by

$$f:A \to B, \quad \text{or} \quad A \xrightarrow{f} B. \tag{5}$$

If x is an element of the set A, the element of B which is assigned to x by the mapping f is denoted by $f(x)$ and is called the image of x. This can conveniently be pictured with the help of the diagram (Fig. 1.4).

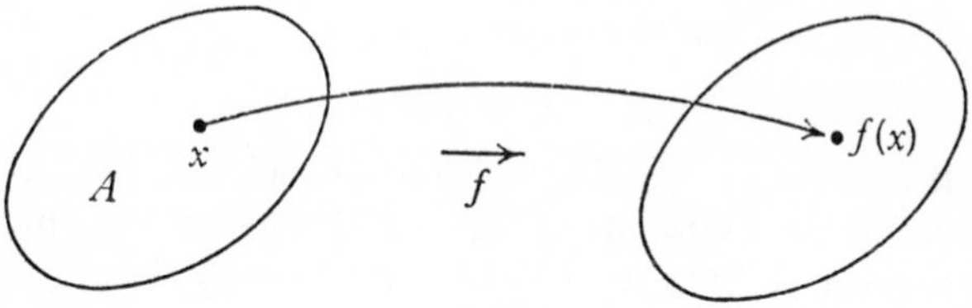

Fig. 1.4

A special mapping is the identity mapping. This is denoted by $f:A \to A$ and sends each element x of A into itself. In other words, $f(x) = x$ (i.e. x is its own image). It is usual to denote the identity mapping more compactly by I.

We now give two examples of simple mappings.

(a) If A is the set of real numbers x, and if f assigns to each number its exponential, then $f(x) = e^x$ are the elements of B, B being the set of positive real numbers.

(b) Let A be the set of the twenty-six letters of the alphabet. If f denotes the mapping which assigns to the first letter, a, the number 1, to b the number 2, and so on so that the last letter z is assigned the number 26, then we may write

$$f \equiv \begin{Bmatrix} a \to 1 \\ b \to 2 \\ c \to 3 \\ \cdot \quad \cdot \\ \cdot \quad \cdot \\ \cdot \quad \cdot \\ z \to 26 \end{Bmatrix}.$$

The elements of B are the integers $1, 2, 3 \ldots 26$. Both these mappings (transformations, functions) are called one-to-one by which we mean that for every element y of B there is an element x of A such that $f(x) = y$, and that if x and x' are two different elements of A then they have different images in B (i.e. $f(x) \neq f(x')$). Given a one-to-one mapping f an inverse mapping f^{-1} can always be found which undoes the work of f. For if f sends x into y so that $y = f(x)$, and f^{-1} sends y into x so that $x = f^{-1}(y)$, then

$$y = f[f^{-1}(y)] = ff^{-1}(y) \tag{6}$$

and

$$x = f^{-1}[f(x)] = f^{-1}f(x). \tag{7}$$

Hence we have

$$ff^{-1} = f^{-1}f = I, \tag{8}$$

where I is the identity mapping which maps each element onto itself. In example (a) the inverse mapping f^{-1} is clearly that mapping which assigns to each element its logarithm (to base e) since

$$\log_e e^x = x \quad \text{and} \quad e^{\log_e x} = x.$$

The inverse of the product of two or more mappings or transformations (provided they are both one-to-one) can easily be found. For suppose f sends x into y and g sends y into z so that

$$y = f(x) \quad \text{and} \quad z = g(y). \tag{9}$$

Then

$$z = g[f(x)], \tag{10}$$

which, by definition, means first perform f on x and then g on $f(x)$. Consequently

$$x = (gf)^{-1}(z). \tag{11}$$

But from (9) we have

$$x = f^{-1}(y) \quad \text{and} \quad y = g^{-1}(z). \tag{12}$$

Consequently

$$x = f^{-1}[g^{-1}(z)] = (f^{-1}g^{-1})(z). \tag{13}$$

Comparing (11) and (13) we find

$$(gf)^{-1} = f^{-1}g^{-1}. \tag{14}$$

The inverse of the product of two one-to-one transformations is obtained therefore by carrying out the inverse transformations one-by-one in reverse order.

One-to-one mappings are frequently used in setting up codes. For example, the mapping of the alphabet onto itself shifted four positions to the left as shown

$$\begin{array}{cccccccccccccc} a & b & c & d & e & \ldots & s & t & u & v & w & x & y & z \\ \updownarrow & \updownarrow & \updownarrow & \updownarrow & \updownarrow & & \updownarrow & \updownarrow & \updownarrow & \updownarrow & \updownarrow & \updownarrow & \updownarrow & \updownarrow \\ e & f & g & h & i & \ldots & w & x & y & z & a & b & c & d \end{array}$$

transforms '*set theory*' into '*wix xlisvc*'.

Not all mappings are one-to-one. For example, the mapping f defined by Fig. 1.5, where x is the image of a, and z is the image of

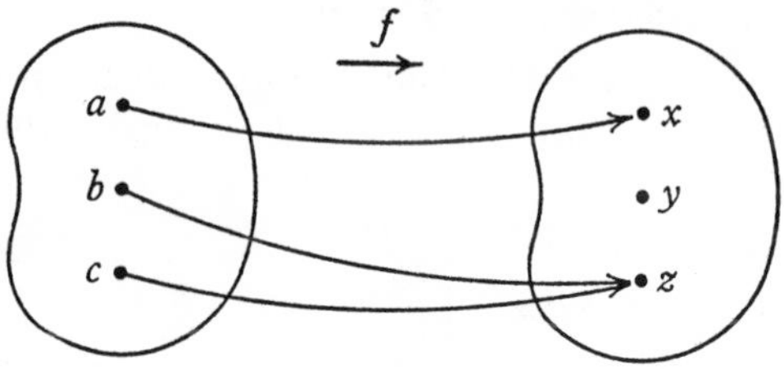

Fig. 1.5

both b and c, does not have an inverse mapping, although of course, inverses of the individual elements exist; these are $f^{-1}(x) = a$, $f^{-1}(z) = \{b, c\}$ (i.e. the set containing the two elements b and c), and $f^{-1}(y) = \emptyset$ (the null set) since neither a, b nor c is mapped into y.

It is clear that if f, g and h are any three mappings then

$$f\{g[h(x)]\} = (fg)[h(x)] = fgh(x) = f(gh)(x) \tag{15}$$

– that is, that the associative law is true. However, it is not true that two mappings necessarily commute – that is, that the product is independent of the order in which the mappings are carried out. For suppose

$$f = \left\{ \begin{matrix} a \to b \\ b \to c \\ c \to a \end{matrix} \right\} \quad \text{and} \quad g = \left\{ \begin{matrix} a \to b \\ b \to b \\ c \to a \end{matrix} \right\}. \tag{16}$$

If we first carry out the mapping g and then the mapping f we find

$$fg = \left\{ \begin{matrix} a \to c \\ b \to c \\ c \to b \end{matrix} \right\}. \tag{17}$$

Conversely, carrying out first f and then g we find

$$gf = \left\{ \begin{matrix} a \to b \\ b \to a \\ c \to b \end{matrix} \right\}. \tag{18}$$

Clearly $fg \neq gf$, showing that f and g do not commute. It might be suspected that non-commutation arises in this particular instance since f is a one-to-one mapping whilst g is not. However, even two one-to-one mappings do not necessarily commute. Nevertheless, two mappings which always commute are a one-to-one mapping f and its inverse f^{-1} (see (8)).

1.5 Linear transformations and matrices

Consider now the two-dimensional problem of the rotation of rectangular Cartesian axes $x_1 0 x_2$ through an angle θ into $y_1 0 y_2$ (see Fig. 1.6).

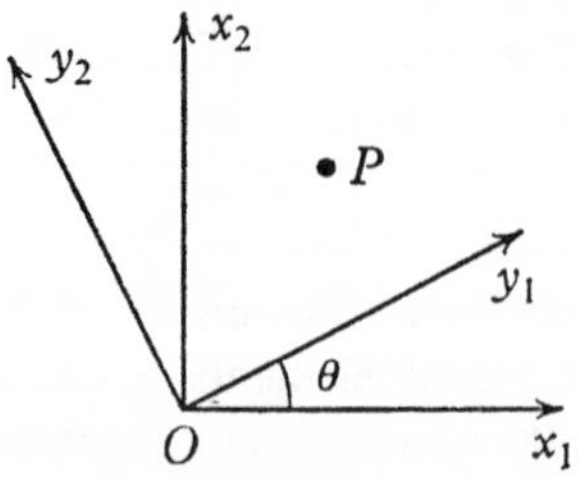

Fig. 1.6

If P is a typical point in the plane of the axes then its coordinates (y_1, y_2) with respect to the $y_1 0 y_2$ system are easily found to be related to its coordinates (x_1, x_2) with respect to the $x_1 0 x_2$ system by the relations

$$\left.\begin{aligned} y_1 &= x_1 \cos\theta + x_2 \sin\theta, \\ y_2 &= -x_1 \sin\theta + x_2 \cos\theta. \end{aligned}\right\} \tag{19}$$

These equations define a mapping of the x_1x_2-plane onto the y_1y_2-plane and form a simple example of a linear transformation. The general linear transformation is defined by the equations

$$\left.\begin{array}{l} y_1 = a_{11}x_1 + a_{12}x_2 + \ . \quad . \quad . + a_{1n}x_n, \\ y_2 = a_{21}x_1 + a_{22}x_2 + \ . \quad . \quad . + a_{2n}x_n, \\ \quad \cdot \qquad \cdot \qquad \cdot \qquad\qquad\qquad \cdot \\ \quad \cdot \qquad \cdot \qquad \cdot \qquad\qquad\qquad \cdot \\ \quad \cdot \qquad \cdot \qquad \cdot \qquad\qquad\qquad \cdot \\ y_m = a_{m1}x_1 + a_{m2}x_2 + \ . \quad . \quad . + a_{mn}x_n. \end{array}\right\} \tag{20}$$

in which the set of n quantities $(x_1, x_2, x_3, \ldots, x_n)$ (the coordinates of a point in an n-dimensional space, say) are transformed linearly into the set of m quantities $(y_1, y_2, \ldots, y_m)$ (the coordinates of a point in an m-dimensional space). This set of equations may be written more concisely as

$$y_i = \sum_{k=1}^{n} a_{ik}x_k \quad (i = 1, 2, \ldots, m), \tag{21}$$

or, in symbolic form, as

$$\mathbf{Y} = \mathbf{A}\mathbf{X}, \tag{22}$$

where

$$\mathbf{Y} = \begin{pmatrix} y_1 \\ y_2 \\ \cdot \\ \cdot \\ \cdot \\ y_m \end{pmatrix}, \quad \mathbf{X} = \begin{pmatrix} x_1 \\ x_2 \\ \cdot \\ \cdot \\ \cdot \\ x_n \end{pmatrix} \quad \text{and} \quad \mathbf{A} = \begin{pmatrix} a_{11} & a_{12} & \cdot & \cdot & \cdot & a_{1n} \\ a_{21} & a_{22} & \cdot & \cdot & \cdot & a_{2n} \\ \cdot & \cdot & & & & \cdot \\ \cdot & \cdot & & & & \cdot \\ \cdot & \cdot & & & & \cdot \\ a_{m1} & a_{m2} & \cdot & \cdot & \cdot & a_{mn} \end{pmatrix}$$

The rectangular array, $\mathbf{A}$, of mn quantities arranged in m rows and n columns is called a matrix of order $(m \times n)$ and must be thought of as operating on $\mathbf{X}$ in such a way as to reproduce the right-hand side of (20). The quantities a_{ik} are called the elements of the matrix $\mathbf{A}$, a_{ik} being the element in the ith row and k^{th} column. We now see that

Y and **X** are matrices of order $(m \times 1)$ and $(n \times 1)$ respectively – matrices having just one column, such as these, are called column matrices.

Of particular importance are square matrices which have the same number of rows as columns (order $(m \times m)$). A simple example of the occurrence of a square matrix is given by writing (19) in symbolic form

$$\mathbf{Y} = \mathbf{AX}, \tag{23}$$

where

$$\mathbf{Y} = \begin{pmatrix} y_1 \\ y_2 \end{pmatrix}, \quad \mathbf{X} = \begin{pmatrix} x_1 \\ x_2 \end{pmatrix} \quad \text{and} \quad \mathbf{A} = \begin{pmatrix} \cos\theta & \sin\theta \\ -\sin\theta & \cos\theta \end{pmatrix}. \tag{24}$$

Here **A** is a (2×2) matrix which operates on **X** to produce **Y**.

Now clearly the general linear transformation (20) with $m \neq n$ cannot be a one-to-one transformation since the number of elements in the set $(x_1, x_2, \ldots, x_n)$ is different from the number in the set $(y_1, y_2, \ldots, y_m)$. An inverse transformation to (20) cannot exist therefore, and consequently we should not expect to be able to find an inverse matrix $\mathbf{A}^{-1}$ (say) which undoes the work of **A**. Indeed, inverses of non-square matrices are not defined. However, if $m = n$ it may be possible to find an inverse transformation and an associated inverse matrix. Consider, for example, the transformation (19). Solving these equations for x_1 and x_2 using Cramer's rule† we have

$$x_1 = \frac{\begin{vmatrix} y_1 & \sin\theta \\ y_2 & \cos\theta \end{vmatrix}}{\begin{vmatrix} \cos\theta & \sin\theta \\ -\sin\theta & \cos\theta \end{vmatrix}}, \quad x_2 = \frac{\begin{vmatrix} \cos\theta & y_1 \\ -\sin\theta & y_2 \end{vmatrix}}{\begin{vmatrix} \cos\theta & \sin\theta \\ -\sin\theta & \cos\theta \end{vmatrix}}. \tag{25}$$

Consequently unique values of x_1 and x_2 exist since the determinant in the denominators of (25) is non-zero. This determinant is in fact just the determinant of the square matrix **A** in (24). In general, an inverse transformation exists provided the determinant of the square matrix inducing the transformation does not vanish. Matrices with non-zero determinants are called non-singular – otherwise they are singular. In Chapter 3 we discuss in detail how to construct the inverse of a non-singular matrix. However, again using our knowledge of mappings we can anticipate one result which will be proved

† See, for example, the author's *Mathematical Methods forScience Students*. Longman, 2nd edition 1973 (Chapter 16).

later (see Chapter 3, 3.4) – namely, that if **A** and **B** are two non-singular matrices inducing linear transformations (mappings) then (cf. equation (14)) the inverse of the product **AB** is given by

$$(\mathbf{AB})^{-1} = \mathbf{B}^{-1}\mathbf{A}^{-1}. \tag{26}$$

Consider now two linear transformations

$$z_i = \sum_{k=1}^{n} b_{ik} y_k \quad \text{with} \quad i = 1, 2, \ldots, m, \tag{27}$$

and

$$y_k = \sum_{j=1}^{p} a_{kj} x_j \quad \text{with} \quad k = 1, 2, \ldots, n. \tag{28}$$

In symbolic form these may be written as

$$\mathbf{Z} = \mathbf{BY} \quad \text{and} \quad \mathbf{Y} = \mathbf{AX}, \tag{29}$$

where

$$= \begin{pmatrix} z_1 \\ z_2 \\ \cdot \\ \cdot \\ \cdot \\ z_m \end{pmatrix}, \qquad \mathbf{Y} = \begin{pmatrix} y_1 \\ y_2 \\ \cdot \\ \cdot \\ \cdot \\ y_n \end{pmatrix}, \qquad \mathbf{X} = \begin{pmatrix} x_1 \\ x_2 \\ \cdot \\ \cdot \\ \cdot \\ x_p \end{pmatrix} \tag{30}$$

and

$$\mathbf{A} = \begin{pmatrix} a_{11} & a_{12} & \cdots & a_{1p} \\ a_{21} & a_{22} & \cdots & a_{2p} \\ \cdot & \cdot & & \cdot \\ \cdot & \cdot & & \cdot \\ \cdot & \cdot & & \cdot \\ a_{n1} & a_{n2} & \cdots & a_{np} \end{pmatrix}, \qquad \mathbf{B} = \begin{pmatrix} b_{11} & b_{12} & \cdots & b_{1n} \\ b_{21} & b_{22} & \cdots & b_{2n} \\ \cdot & \cdot & & \cdot \\ \cdot & \cdot & & \cdot \\ \cdot & \cdot & & \cdot \\ b_{m1} & b_{m2} & \cdots & b_{mn} \end{pmatrix}. \tag{31}$$

The result of first transforming $(x_1, x_2, \ldots, x_p)$ into $(y_1, y_2, \ldots, y_n)$ by (28), and then transforming $(y_1, y_2, \ldots, y_n)$ into $(z_1, z_2, \ldots, z_m)$ by (27) is given by

$$z_i = \sum_{k=1}^{n} b_{ik} \sum_{j=1}^{p} a_{kj} x_j = \sum_{j=1}^{p} \left(\sum_{k=1}^{n} b_{ik} a_{kj} \right) x_j. \tag{32}$$

Symbolically this is equivalent to

$$\mathbf{Z} = \mathbf{BY} = \mathbf{BAX}, \tag{33}$$

where the operator **BA** must be thought of as transforming **X** into **Z**. Suppose, however, we go direct from $(x_1, x_2, \ldots, x_p)$ into

$(z_1, z_2, \ldots, z_m)$ by the transformation

$$z_i = \sum_{j=1}^{p} c_{ij} x_j, \tag{34}$$

which in symbolic form reads

$$\mathbf{Z} = \mathbf{CX}, \tag{35}$$

where

$$\mathbf{C} = \begin{pmatrix} c_{11} & c_{12} & \cdots & c_{1p} \\ c_{21} & c_{22} & \cdots & c_{2p} \\ \cdot & \cdot & & \cdot \\ \cdot & \cdot & & \cdot \\ \cdot & \cdot & & \cdot \\ c_{m1} & c_{m2} & \cdots & c_{mp} \end{pmatrix}. \tag{36}$$

Then comparing (34) and (32) we find

$$c_{ij} = \sum_{k=1}^{n} b_{ik} a_{kj}. \tag{37}$$

Equation (37) gives the elements c_{ik} of the matrix **C** in terms of the elements a_{ik} of **A** and b_{ik} of **B**. However, from (35) and (33) we see that **C** = **BA** so (37) in fact gives the elements of the matrix product **BA**. Clearly for this product to exist the number of columns of **B** must be equal to the number of rows of **A** (see (37) where the summation is on the columns of **B** and the rows of **A**). The order of the resulting matrix **C** is $(m \times p)$.

As an example of the product of two matrices we can justify the earlier statement **Y** = **AX** (see equation (23)) since, using (37),

$$\mathbf{AX} = \begin{pmatrix} \cos\theta & \sin\theta \\ -\sin\theta & \cos\theta \end{pmatrix} \begin{pmatrix} x_1 \\ x_2 \end{pmatrix}$$

$$= \begin{pmatrix} x_1 \cos\theta + x_2 \sin\theta \\ -x_1 \sin\theta + x_2 \cos\theta \end{pmatrix} = \begin{pmatrix} y_1 \\ y_2 \end{pmatrix} = \mathbf{Y}. \tag{38}$$

Our aim so far has been to show the close relationship between linear transformations and matrices. In Chapter 2, 2.1, matrix multiplication and other matrix operations will be dealt with in greater detail and in a more formal way.

1.6 Occurrence and uses of matrices

Although matrices were first introduced in 1857 by Cayley, it was not until the early 1920s when Heisenberg, Born and others realised their use in the development of quantum theory that matrices became of interest to physicists. Nowadays, matrices are of interest and use to mathematicians, scientists and engineers alike, occurring as they do in such a variety of subjects as electric circuit theory, oscillations, wave propagation, quantum mechanics, field theory, atomic and molecular structure – as well as being a most powerful tool in many parts of mathematics such as the stability of differential equations, group theory, difference equations and computing. In the fields of probability and statistics, game theory, and mathematical economics, matrices are also widely used.

It is instructive to give at this stage a simple illustration of the formulation of a physical problem in matrix language.

Consider the problem of the small vertical oscillations of two masses m_1 and m_2 attached to two massless springs of stiffness s_1 and

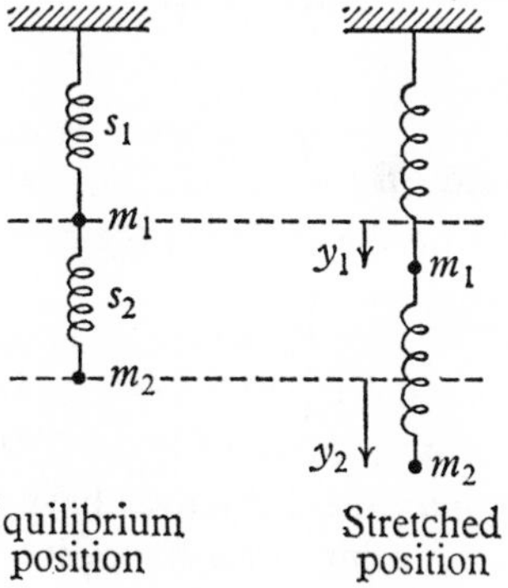

Fig. 1.7

s_2 (see Fig. 1.7). If y_1 and y_2 are the displacements from the equilibrium position at time t, the equations of motion are

$$\left.\begin{aligned} m_1\ddot{y}_1 &= -s_1 y_1 + s_2(y_2 - y_1), \\ m_2\ddot{y}_2 &= -s_2(y_2 - y_1), \end{aligned}\right\} \tag{39}$$

(dots denoting differentiation with respect to time). These equations may be written in matrix form as

$$\ddot{\mathbf{Y}} = \mathbf{AY}, \tag{40}$$

where

$$\mathbf{Y} = \begin{pmatrix} y_1 \\ y_2 \end{pmatrix}, \quad \text{and} \quad \mathbf{A} = \begin{pmatrix} -\dfrac{(s_1+s_2)}{m_1} & \dfrac{s_2}{m_1} \\ \dfrac{s_2}{m_2} & -\dfrac{s_2}{m_2} \end{pmatrix}. \tag{41}$$

If desired the second time derivatives may be eliminated from (39) by the introduction of two new dependent variables y_3 and y_4 such that

$$\left.\begin{aligned} m_1 \dot{y}_1 &= y_3, \\ m_2 \dot{y}_2 &= y_4. \end{aligned}\right\} \tag{42}$$

Then

$$\left.\begin{aligned} \dot{y}_3 &= -s_1 y_1 + s_2(y_2 - y_1), \\ \dot{y}_4 &= -s_2(y_2 - y_1). \end{aligned}\right\} \tag{43}$$

In matrix form (42) and (43) read

$$\dot{\mathbf{W}} = \mathbf{BW}, \tag{44}$$

where

$$\mathbf{W} = \begin{pmatrix} y_1 \\ y_2 \\ y_3 \\ y_4 \end{pmatrix} \quad \text{and} \quad \mathbf{B} = \begin{pmatrix} 0 & 0 & \dfrac{1}{m_1} & 0 \\ 0 & 0 & 0 & \dfrac{1}{m_2} \\ -(s_1+s_2) & s_2 & 0 & 0 \\ s_2 & -s_2 & 0 & 0 \end{pmatrix}. \tag{45}$$

We see that the second time derivatives have been eliminated only at the expense of introducing larger matrices.

1.7 Operations with sets

In the earlier sections of this chapter we introduced the ideas of set theory and developed the set notation just far enough to enable the concept of a mapping to be understood. Of course, it would have been possible to omit the set theory sections and to introduce matrices just by linear transformations. However, as mentioned earlier, set theory is becoming increasingly used in many branches of science and engineering; consequently having already introduced some of its basic notions we follow it a little farther here.

A common criticism of introducing set theory to scientists and engineers (and for that matter to school children, as is now fashionable) is that it is only notation and that little or nothing can be done using set formalism that cannot be done in a more conventional way. Although to some extent this may be true it is equally true of a large part of mathematics as a whole; the development of a new notation often has a unifying and simplifying effect and suggests lines of further development. For example, it is more convenient to deal with the Arabic numbers rather than the clumsy Roman form; vectors are more convenient in many cases than dealing separately with their components, and linear transformations are better dealt with in matrix form than by writing down a set of n linear equations.

We shall not pursue set theory very far, but will go just a sufficient distance to show some of the types of problems in which sets may be used with profit.

In what follows we let U denote the set of elements under discussion (the universal set) and A and B be two subsets of U.

(a) *Union and intersection of sets*

The union (or join or logical sum) of A and B is denoted by $A \cup B$ and is defined as the set of all elements belonging to either A or B or both (see, for example, the shaded part of the Venn diagram in Fig. 1.8). The symbol $\cup$ is usually read as ' cup '.

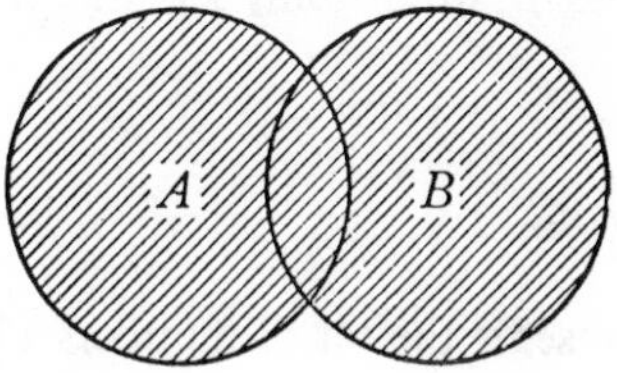

$A \cup B$ represented by shaded region

Fig. 1.8

Clearly

$$A \cup B = B \cup A, \tag{46}$$

and, since A and B are subsets of $A \cup B$,

$$A \subseteq (A \cup B), \quad \text{and} \quad B \subseteq (A \cup B). \tag{47}$$

Likewise

$$A \cup U = U \cup A = U. \tag{48}$$

The intersection (or meet or logical product) of A and B is denoted by $A \cap B$ and is the set of those elements common to both A and B (see the shaded part of the Venn diagram in Fig. 1.9). The symbol $\cap$ is read as ' cap '.

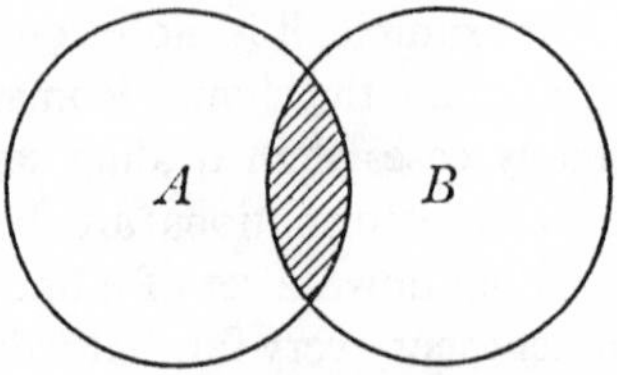

$A \cap B$ represented by shaded region

Fig. 1.9

Corresponding results to those for the union are

$$A \cap B = B \cap A, \tag{49}$$

$$(A \cap B) \subseteq A, \qquad (A \cap B) \subseteq B, \tag{50}$$

and

$$A \cap U = U \cap A = A. \tag{51}$$

Furthermore, if A and B are disjoint sets (i.e. no elements in common) then

$$A \cap B = \emptyset, \tag{52}$$

where $\emptyset$ is the empty set.

Example 1. If A represents the set of numbers $\{1, 2, 3, 4, 5, 6\}$, and B the set of numbers $\{5, 6, 7\}$, then

$$A \cup B = \{1, 2, 3, 4, 5, 6, 7\} \tag{53}$$

and

$$A \cap B = \{5, 6\}. \tag{54}$$

Example 2. Let A be the set of points in the region of the Euclidean plane defined by $|x| \leqslant 1$, $|y| \leqslant 1$, and B the set of points in the region defined by $y^2 \leqslant x, 0 \leqslant x \leqslant 2$. The sets $A \cup B$ and $A \cap B$ are then represented by the shaded parts of Figs. 1.10 and 1.11 respectively.

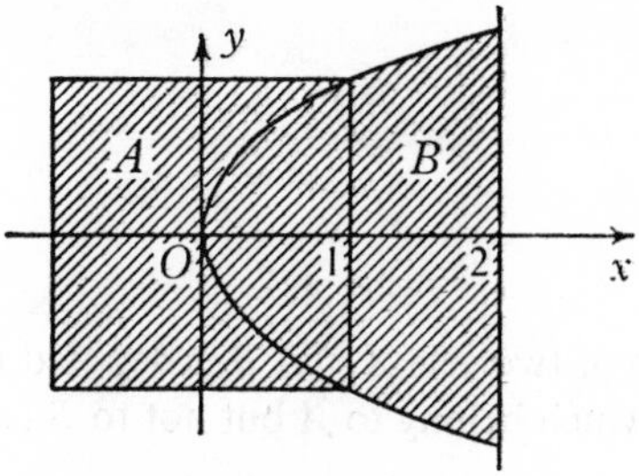

$A \cup B$ represented by shaded region

Fig. 1.10

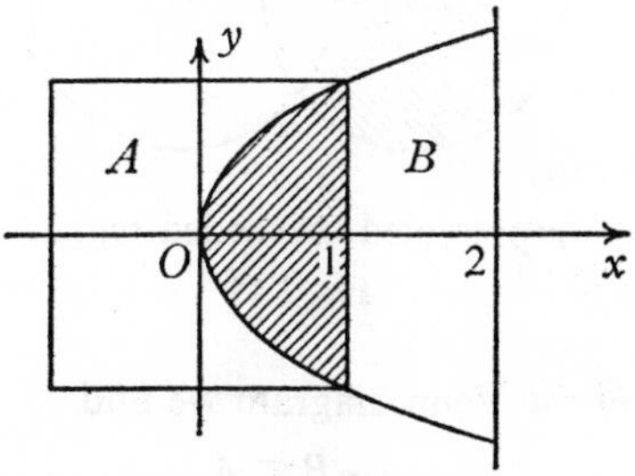

$A \cap B$ represented by shaded region

Fig 1.11

(b) *Complement of a set and difference of sets*

The complement of a set A is denoted by A' and is the set of elements which do not belong to A. Accordingly if A is a subset of the universal set U (represented by the rectangle in Fig. 1.12), then A' is represented by the shaded part of the diagram.

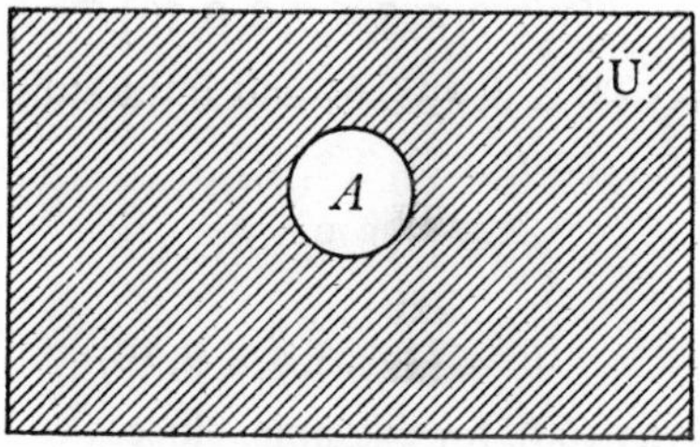

A' represented by shaded region

Fig. 1.12

It is clear that

$$(A')' = A, \tag{55}$$

$$A \cup A' = U, \tag{56}$$

and

$$A \cap A' = \emptyset. \tag{57}$$

The difference of two sets A and B is denoted by $A-B$ and is the set of elements which belong to A but not to B (see Fig. 1.13).

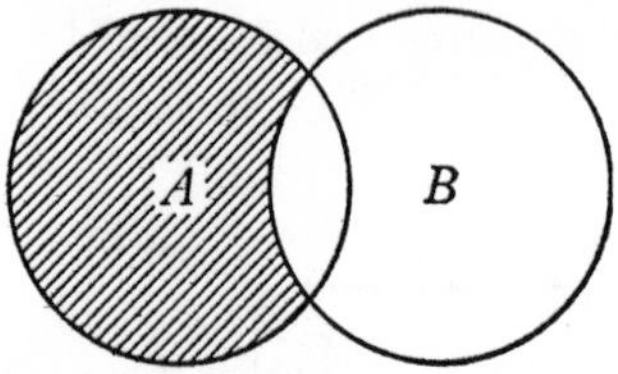

$A-B$
represented by shaded region
Fig. 1.13

By inspection of the Venn diagram we find

$$A-B \subseteq A, \tag{58}$$

$$A \cup B = (A-B) \cup B, \tag{59}$$

and

$$A-B = A \cap B'. \tag{60}$$

Furthermore, $A-B$, $A \cap B$ and $B-A$ are disjoint sets. Hence

$$(A-B) \cap (A \cap B) = \emptyset, \tag{61}$$

and so on.

Example 3. Suppose U is the set of numbers

$$\{1,2,3,4,5,6,7,8,9,10\}.$$

Let A be the subset $\{1, 2, 3, 4, 5, 6\}$ and B the subset $\{5, 6, 7\}$. Then

$$A-B = \{1,2,3,4\}, \tag{62}$$

$$A' = \{7,8,9,10\}, \tag{63}$$

and

$$B' = \{1,2,3,4,8,9,10\}. \tag{64}$$

We may easily verify (60), for example, since

$$A \cap B' = \{1,2,3,4\}, \tag{65}$$

which is identical with (62).

1.8 Set algebra

It will have been noticed in the previous section that various relationships hold between the four operations $\cup$, $\cap$, $-$, and $'$. These are in fact just examples of the laws of set algebra, the most important of which we give here. In these relations A, B and C are subsets of the universal set U.

(a) $$\left.\begin{aligned} &U' = \emptyset, \quad A \cap A' = \emptyset, \\ &\emptyset' = U, \quad A \cup A' = U, \quad (A')' = A. \end{aligned}\right\} \qquad (66)$$

(b) $$A \cup A = A, \qquad A \cap A = A. \qquad (67)$$

(c) $$\left.\begin{aligned} &A \cup U = U, \quad A \cup \emptyset = A, \quad A - \emptyset = A, \\ &A \cap U = A, \quad A \cap \emptyset = \emptyset, \quad A - A = \emptyset. \end{aligned}\right\} \qquad (68)$$

(d) $$\begin{aligned} A \cap B = B \cap A \\ A \cup B = B \cup A \end{aligned} \quad \text{(commutative laws).} \qquad (69)$$

(e) $$\begin{aligned} (A \cup B) \cup C = A \cup (B \cup C) \\ (A \cap B) \cap C = A \cap (B \cap C) \end{aligned} \quad \text{(associative laws).} \qquad (70)$$

(f) $$\begin{aligned} A \cup (B \cap C) = (A \cup B) \cap (A \cup C) \\ A \cap (B \cup C) = (A \cap B) \cup (A \cap C) \end{aligned} \quad \text{(distributive laws).} \qquad (71)$$

These relations may be easily verified by Venn diagrams. For example, in Fig. 1.14a the horizontally shaded part represents $A \cap B$, and the horizontally and vertically shaded part $(A \cap B) \cap C$.

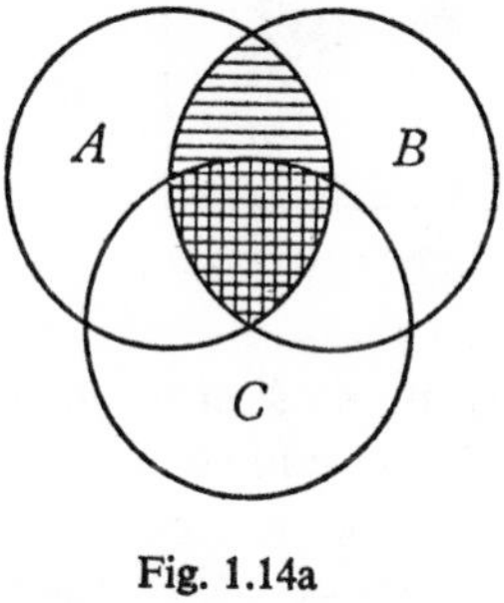

Fig. 1.14a

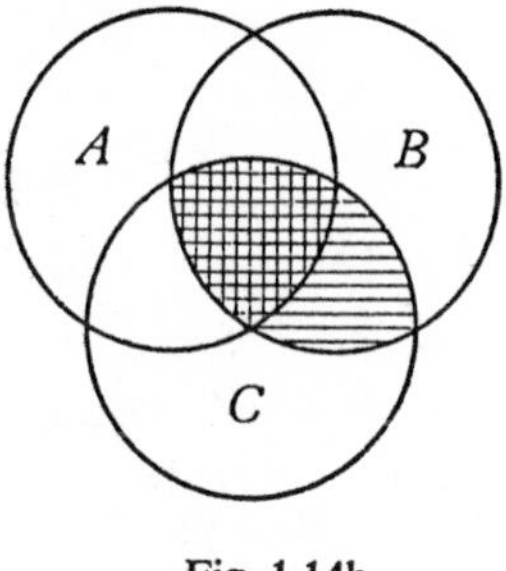

Fig. 1.14b

Likewise in Fig. 1.14b, the horizontally shaded part represents $B \cap C$, and the horizontally and vertically shaded part $A \cap (B \cap C)$.

Clearly $(A \cap B) \cap C = A \cap (B \cap C)$, as stated in (70).

1.9 Some elementary applications of set theory

It is impossible here to give an overall picture of the applications of set theory. Indeed, much of the importance of set theory lies in the more abstract and formal branches of pure mathematics. However, the following examples give some indication of a few of the problems which may be dealt with using sets.

Example 4. In a survey of 100 students it was found that 40 studied mathematics, 64 studied physics, 35 studied chemistry, 1 studied all three subjects, 25 studied mathematics and physics, 3 studied mathematics and chemistry, and 20 studied physics and chemistry. To find the number who studied chemistry only, and the number who studied none of these subjects.

Here the basic set under discussion U (the universal set) is the set of 100 students in the survey. This set is represented in the usual way by a rectangle (see Fig. 1.15). Let the three overlapping circular

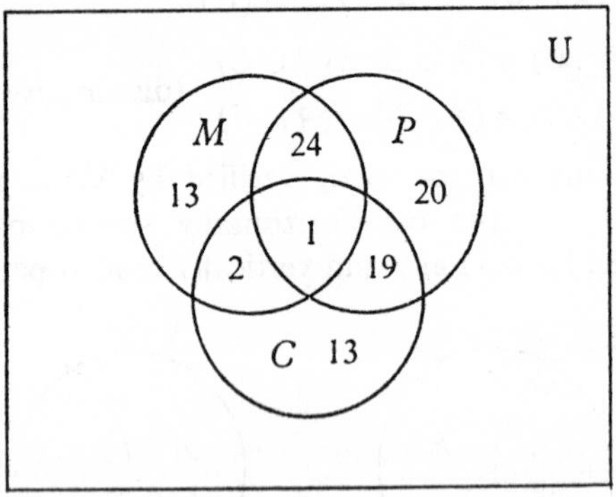

Fig. 1.15

regions M, P and C represent the subsets of U corresponding to those students studying mathematics, physics and chemistry respectively. We see that the intersection of all three subsets $M \cap (P \cap C)$ represents 1 student (and is so labelled). Likewise, since the number of students studying mathematics and chemistry $(M \cap C)$ is 3, the number of students studying only mathematics and chemistry is

$$M \cap C - M \cap (P \cap C) = 3 - 1 = 2. \qquad (2)$$

In this way every part of the Venn diagram may be labelled with the appropriate number of elements. From Fig. 1.15 we see that the numbers of students studying only mathematics, only physics and only chemistry are respectively 13, 20 and 13. Furthermore, the total of the numbers in the subset $(M \cup P) \cup C$ is seen to be 92. Hence the number of students not studying any of the three subjects is

$$[(M \cup P) \cup C]' = U - [(M \cup P) \cup C] = 100 - 92 = 8. \qquad (73)$$

Example 5. The results of surveys are not always consistent. Consistency may be readily checked using Venn diagrams. Suppose out of 900 students it was reported that 700 drove cars, 400 rode bicycles, and 150 both drove cars and rode bicycles. If A represents the set of car-driving students, and B the set of cyclists then $A \cap B = 150$. Hence $A - B = 550$ and $B - A = 250$ (see Fig. 1.16). Since the basic

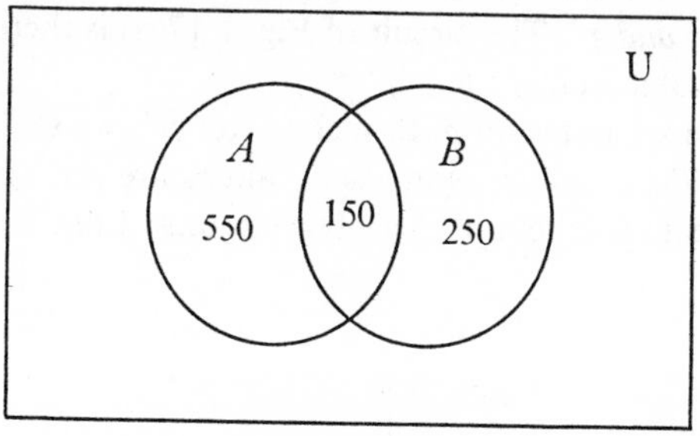

Fig. 1.16

set U contains 900 students and, by inspection,

$$A \cup B = 550 + 250 + 150 = 950,$$

we see that the data must be inconsistent. Put another way, the number of students who neither drive cars nor ride cycles is

$$U - (A \cup B) = 900 - 950 = -50. \qquad (\ 4)$$

Clearly the necessary and sufficient condition for data to be consistent is that the number of elements in each subset must be non-negative.

Example 6. Closely connected with set theory is Boolean algebra. This is an algebraic structure which has laws similar to those of sets (see 1.8). Its importance chiefly lies in the description and design of electrical switching circuits and computing systems. Although it is not possible to give a detailed account of Boolean algebra here, a few simple ideas can be indicated.

Consider the simple circuit shown in Fig. 1.17(a) in which s_1 and s_2 are switches in series. If p denotes the statement 'switch s_1 is open' and q the statement 'switch s_2 is open', then (a) is described by the statement '*p and q*'. In set theory notation we have seen that the intersection $A \cap B$ of two sets A and B defines those elements

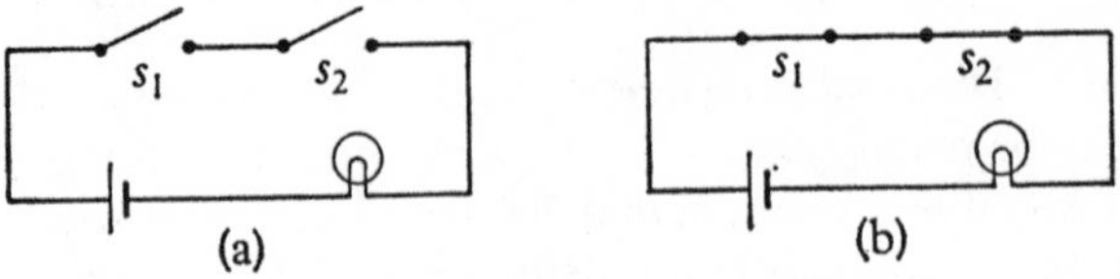

Fig. 1.17

common to *A and B*. Hence we take over the set notation and write $p \cap q$ for '*p and q*'. The circuit of Fig. 1.17(a) is therefore described by the logical statement $p \cap q$.

Using the set notation p' to mean 'not p', we see that the circuit of Fig. 1.17(b) in which switches s_1 and s_2 are not open is described by the logical statement $p' \cap q'$. The circuits of Fig. 1.18 are similarly described.

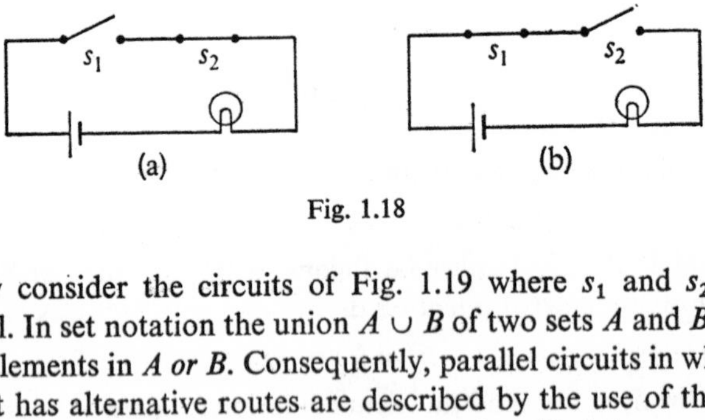

Fig. 1.18

Now consider the circuits of Fig. 1.19 where s_1 and s_2 are in parallel. In set notation the union $A \cup B$ of two sets A and B defines those elements in *A or B*. Consequently, parallel circuits in which the current has alternative routes are described by the use of the union symbol $\cup$. For example, Fig. 1.19(a) is described by $p \cup q'$, Fig. 1.19(b) by $p' \cup q'$, Fig. 1.19(c) by $p \cup q$, and Fig. 1.19(d) by $p' \cup q$.

The description of more complicated circuits can readily be found by treating them as combinations of these basic series and parallel circuits.

Boolean algebra is useful in showing the equivalence of two circuits. For suppose a circuit is described by $(p \cup q) \cap (p \cup r)$. Then since $(p \cup q) \cap (q \cup r) = p \cup (q \cap r)$ is a law of Boolean algebra

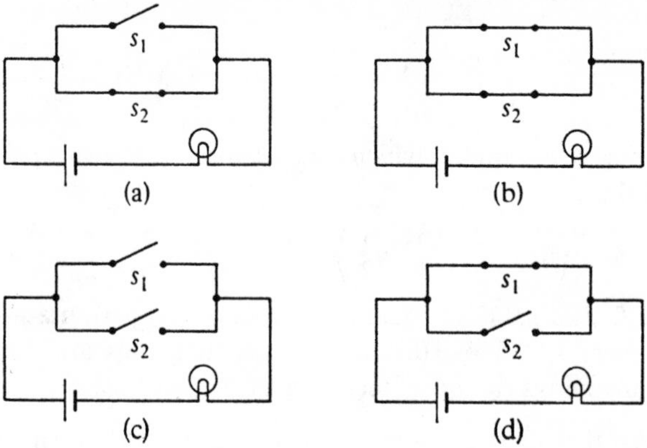

Fig. 1.19

(cf. equation (71)) the circuit must be equivalent to another with the structure $p \cup (q \cap r)$. This brief sketch of Boolean algebra has been included here only to indicate one of the developments of set theory. For further details the reader should consult a more specialised text.

PROBLEMS 1

1. Express in words the statements

   (a) $A = \{x;\, x^2+x-12 = 0\}$,
   (b) $B = \{x;\, \tan x = 0\}$.

   Which of these two sets is finite?

2. Which of the following sets is the null set $\varnothing$?

   (a) $A = \{x;\, x \text{ is} > 1 \text{ and } x \text{ is} < 1\}$,
   (b) $B = \{x;\, x+3 = 3\}$,
   (c) $C = \{\varnothing\}$.

3. (a) If $A = \{1, 2, 3, 4\}$, enumerate all the subsets of A.
   (b) If $B = \{1, \{2, 3\}\}$, enumerate all the subsets of B.

4. Which of the following sets are equal?

   (a) $\{x;\, x \text{ is a positive integer} \leqslant 4\}$,
   (b) $\{1, 2, 3\}$,
   (c) $\{x;\, x \text{ is a prime number} < 5\}$,

(d) $\{1, \{2, 3\}\}$,
(e) $\{1, 2, 3, 1\}$.

5. The symbol

$$\begin{pmatrix} 1 & 2 & 3 & 4 \\ a_1 & a_2 & a_3 & a_4 \end{pmatrix}$$

denotes the mapping which sends 1 into a_1, 2 into a_2, 3 into a_3 and 4 into a_4.

If $f = \begin{pmatrix} 1 & 2 & 3 & 4 \\ 4 & 1 & 3 & 2 \end{pmatrix}$ and $g = \begin{pmatrix} 1 & 2 & 3 & 4 \\ 3 & 2 & 1 & 4 \end{pmatrix}$

find fg and gf. Each mapping (f and g) is a permutation of the numbers 1, 2, 3, 4. How many such mappings are there? Are the mappings fg and gf members of this set?

6. Given that the letter *e* is the most frequently occurring letter in the *decoded* form of the message

gqk xeja bacqbab lxwm sammeya,

obtain the mapping of the alphabet onto itself which decodes the message. The only operations allowed are lateral displacements of the alphabet as a whole, and turning the alphabet backwards.

7. Express the transformation

$$\begin{aligned} y_1 &= 6x_1 + 2x_2 - x_3, \\ y_2 &= x_1 - x_2 + 2x_3, \\ y_3 &= 7x_1 + x_2 + x_3, \end{aligned}$$

in the symbolic form $\mathbf{Y} = \mathbf{AX}$. Determine whether or not an inverse transformation exists.

8. Evaluate the matrix product

$$\begin{pmatrix} 2 & 3 & 1 \\ 3 & 1 & 2 \\ 1 & 2 & 3 \end{pmatrix} \begin{pmatrix} 3 & -1 \\ 1 & 0 \\ -2 & 1 \end{pmatrix}.$$

9. Given $A = \{1, 2, 3, 4\}$, $B = \{3, 4, 5\}$ and $C = \{1, 4, 5\}$, find $A \cup (B \cap C)$, $A \cap (B \cap C)$, $A \cup (B \cap C)$ and $A \cap (B \cup C)$. Verify that $A \cap (B \cup C) = (A \cap B) \cup (A \cap C)$.

10. Verify that $A - (B \cup C) = (A - B) \cap (A - C)$, and that

$$A - (B \cap C) = (A - B) \cup (A - C).$$

CHAPTER 2

Matrix Algebra

2.1 Laws of matrix algebra

In Chapter 1, 1.5 a matrix was defined as an array of mn elements arranged in m rows and n columns. We now consider some of the elementary algebraic operations which may be carried out with matrices, bearing in mind that as with matrix multiplication these may be derived from first principles by appealing to the properties of linear transformations.

(a) *Addition and subtraction of matrices*

The operations of addition and subtraction of matrices are defined only if the matrices which are being added or subtracted are of the same order. If **A** and **B** are two $(m \times n)$ matrices with elements a_{ik} and b_{ik} respectively, then their sum $\mathbf{A}+\mathbf{B}$ is the $(m \times n)$ matrix **C** whose elements c_{ik} are given by $c_{ik} = a_{ik}+b_{ik}$. Likewise $\mathbf{A}-\mathbf{B}$ is the $(m \times n)$ matrix **D** whose elements d_{ik} are given by $d_{ik} = a_{ik}-b_{ik}$. For example, if

$$\mathbf{A} = \begin{pmatrix} 1 & 2 & 3 \\ -1 & 1 & 2 \end{pmatrix} \quad \text{and} \quad \mathbf{B} = \begin{pmatrix} 3 & 4 & 1 \\ 0 & 2 & -1 \end{pmatrix} \tag{1}$$

then

$$\mathbf{A}+\mathbf{B} = \begin{pmatrix} 4 & 6 & 4 \\ -1 & 3 & 1 \end{pmatrix} \quad \text{and} \quad \mathbf{A}-\mathbf{B} = \begin{pmatrix} -2 & -2 & 2 \\ -1 & -1 & 3 \end{pmatrix}. \tag{2}$$

Two matrices are said to be conformable to addition and subtraction if they are of the same order. No meaning is attached to the sum or difference of two matrices of differing orders. From the definition of the addition of matrices it can now be seen that if **A**, **B** and **C** are three matrices conformable to addition then

$$\mathbf{A}+\mathbf{B} = \mathbf{B}+\mathbf{A} \tag{3}$$

and

$$\mathbf{A}+(\mathbf{B}+\mathbf{C}) = (\mathbf{A}+\mathbf{B})+\mathbf{C} = \mathbf{A}+\mathbf{B}+\mathbf{C}. \tag{4}$$

These two results are respectively the commutative law of addition and the associative law of addition.

(b) *Equality of matrices*

Two matrices **A** and **B** with elements a_{ik} and b_{ik} respectively are equal only if they are of the same order and if all their corresponding elements are equal (i.e. if $a_{ik} = b_{ik}$ for all i, k).

(c) *Multiplication of a matrix by a number*

The result of multiplying a matrix **A** (with elements a_{ik}) by a number k (real or complex) is defined as a matrix **B** whose elements b_{ik} are k times the elements of **A**. For example, if

$$\mathbf{A} = \begin{pmatrix} 1 & 2 \\ 3 & 4 \end{pmatrix} \quad \text{then} \quad 6\mathbf{A} = \begin{pmatrix} 6 & 12 \\ 18 & 24 \end{pmatrix}. \tag{5}$$

From this definition it follows that the distributive law

$$k(\mathbf{A}\pm\mathbf{B}) = k\mathbf{A}\pm k\mathbf{B} \tag{6}$$

is valid (provided, of course, that **A** and **B** are conformable to addition). Furthermore, we define

$$k\mathbf{A} = \mathbf{A}k \tag{7}$$

so that multiplication of a matrix by a number is commutative.

(d) *Matrix products*

As we have already seen in Chapter 1, 1.5, two matrices **A** and **B** can be multiplied together to form their product **BA** (in that order) only when the number of columns of **B** is equal to the number of rows of **A**. **A** and **B** are then said to be conformable to the product **BA**. We shall see shortly, however, that **A** and **B** need not be conformable to the product **AB**, and that, even when they are, the product **AB** does not necessarily equal the product **BA**. That is, matrix multiplication is in general non-commutative.

Suppose now **A** is a matrix of order $(m \times p)$ with elements a_{ik}, and **B** is a matrix of order $(p \times n)$ with elements b_{ik}. Then **A** and **B** are conformable to the product **AB** which is a matrix **C**, say, of order $(m \times n)$ with elements c_{ik} defined by

$$c_{ik} = \sum_{s=1}^{p} a_{is} b_{sk}. \tag{8}$$

For example, if **A** and **B** are the matrices

$$\mathbf{A} = \begin{pmatrix} a_{11} & a_{12} \\ a_{21} & a_{22} \\ a_{31} & a_{32} \end{pmatrix}, \quad \mathbf{B} = \begin{pmatrix} b_{11} & b_{12} \\ b_{21} & b_{22} \end{pmatrix} \tag{9}$$

then **C** = **AB** is the (3×2) matrix (using (8))

$$C = \begin{pmatrix} a_{11}b_{11}+a_{12}b_{21} & a_{11}b_{12}+a_{12}b_{22} \\ a_{21}b_{11}+a_{22}b_{21} & a_{21}b_{12}+a_{22}b_{22} \\ a_{31}b_{11}+a_{32}b_{21} & a_{31}b_{12}+a_{32}b_{22} \end{pmatrix}. \tag{10}$$

The product **BA**, however, is not defined since the number of columns of **B** (i.e. two) is not equal to the number of rows of **A** (i.e. three) – in other words, **A** and **B** are not conformable to the product **BA**.

As another example, we take the matrices

$$\mathbf{A} = \begin{pmatrix} 3 & 1 & 2 \\ 2 & 1 & 3 \end{pmatrix}, \qquad \mathbf{B} = \begin{pmatrix} 1 & 2 \\ 3 & 1 \\ 2 & 3 \end{pmatrix}. \tag{11}$$

Then

$$\mathbf{AB} = \begin{pmatrix} 10 & 13 \\ 11 & 14 \end{pmatrix}. \tag{12}$$

Now the product **BA** is also defined in this case since the number of columns of **B** is equal to the number of rows of **A**. However, it is readily found that

$$\mathbf{BA} = \begin{pmatrix} 7 & 3 & 8 \\ 11 & 4 & 9 \\ 12 & 5 & 13 \end{pmatrix}. \tag{13}$$

Clearly **AB** ≠ **BA**, since the orders of the two matrix products are different (see 2.1(b)). This non-commutative property of matrix multiplication may appear even when the two products are defined and are of the same order. To illustrate this we take

$$\mathbf{A} = \begin{pmatrix} 0 & 1 \\ 1 & 1 \end{pmatrix}, \qquad \mathbf{B} = \begin{pmatrix} 0 & -1 \\ 1 & 0 \end{pmatrix}. \tag{14}$$

Then

$$\mathbf{AB} = \begin{pmatrix} 1 & 0 \\ 1 & -1 \end{pmatrix}, \qquad \mathbf{BA} = \begin{pmatrix} -1 & -1 \\ 0 & 1 \end{pmatrix}, \tag{15}$$

whence again **AB** ≠ **BA**. This shows that matrices behave in a different way from numbers by not obeying (in general) the commutative law of multiplication. However, they still obey the associative law of multiplication and the distributive law in that if **A**, **B** and **C** are three matrices for which the various products and sums are defined then

$$(\mathbf{AB})\mathbf{C} = \mathbf{A}(\mathbf{BC}) \tag{16}$$

and

$$(\mathbf{A}+\mathbf{B})\mathbf{C} = \mathbf{AC}+\mathbf{BC}. \tag{17}$$

Matrices **A** and **B** for which **AB** = **BA** are said to commute under multiplication.

2.2 Partitioning of matrices

In dealing with matrices of high order it is often convenient to break down the original matrix into sub-matrices. This is done by inserting horizontal and vertical lines between the elements. The matrix is then said to be partitioned into sub-matrices. For example, the (3×4) matrix

$$\mathbf{A} = \left(\begin{array}{ccc|c} a_{11} & a_{12} & a_{13} & a_{14} \\ a_{21} & a_{22} & a_{23} & a_{24} \\ \hline a_{31} & a_{32} & a_{33} & a_{34} \end{array}\right) \tag{18}$$

is partitioned into sub-matrices $\boldsymbol{\alpha}_{11}$, $\boldsymbol{\alpha}_{12}$, $\boldsymbol{\alpha}_{21}$ and $\boldsymbol{\alpha}_{22}$ by the straight lines, and may be written as

$$\mathbf{A} = \begin{pmatrix} \boldsymbol{\alpha}_{11} & \boldsymbol{\alpha}_{12} \\ \boldsymbol{\alpha}_{21} & \boldsymbol{\alpha}_{22} \end{pmatrix}, \tag{19}$$

where

$$\left.\begin{aligned} \boldsymbol{\alpha}_{11} &= \begin{pmatrix} a_{11} & a_{12} & a_{13} \\ a_{21} & a_{22} & a_{23} \end{pmatrix}, & \boldsymbol{\alpha}_{12} &= \begin{pmatrix} a_{14} \\ a_{24} \end{pmatrix}, \\ \boldsymbol{\alpha}_{21} &= (a_{31} \quad a_{32} \quad a_{33}), & \boldsymbol{\alpha}_{22} &= (a_{34}). \end{aligned}\right\} \tag{20}$$

Now suppose **A** and **B** are two matrices conformable to addition. Then if **A** and **B** are partitioned as

$$\mathbf{A} = \begin{pmatrix} \boldsymbol{\alpha}_{11} & \boldsymbol{\alpha}_{12} \\ \boldsymbol{\alpha}_{21} & \boldsymbol{\alpha}_{22} \end{pmatrix}, \qquad \mathbf{B} = \begin{pmatrix} \boldsymbol{\beta}_{11} & \boldsymbol{\beta}_{12} \\ \boldsymbol{\beta}_{21} & \boldsymbol{\beta}_{22} \end{pmatrix} \tag{21}$$

we have

$$\mathbf{A}+\mathbf{B} = \begin{pmatrix} \boldsymbol{\alpha}_{11}+\boldsymbol{\beta}_{11} & \boldsymbol{\alpha}_{12}+\boldsymbol{\beta}_{12} \\ \boldsymbol{\alpha}_{21}+\boldsymbol{\beta}_{21} & \boldsymbol{\alpha}_{22}+\boldsymbol{\beta}_{22} \end{pmatrix} \tag{22}$$

provided that for each sub-matrix $\boldsymbol{\alpha}_{ik}$ the corresponding sub-matrix $\boldsymbol{\beta}_{ik}$ is of the same order. This will be so provided **A** and **B** are partitioned in precisely the same way. For example,

$$\left(\begin{array}{c|cc} 1 & 4 & 7 \\ \hline 2 & 5 & 8 \\ 3 & 6 & 9 \end{array}\right) + \left(\begin{array}{c|cc} 0 & 3 & 6 \\ \hline 1 & 4 & 7 \\ 2 & 5 & 8 \end{array}\right) = \left(\begin{array}{c|cc} 1 & 7 & 13 \\ \hline 3 & 9 & 15 \\ 5 & 11 & 17 \end{array}\right). \tag{23}$$

We now come to the problem of the multiplication of partitioned matrices. Suppose **A** is a $(m \times p)$ matrix and **B** is a $(p \times n)$ matrix. Then their product $\mathbf{AB} = \mathbf{C}$ is an $(m \times n)$ matrix with elements c_{ik}, where

$$c_{ik} = \sum_{s=1}^{p} a_{is} b_{sk} \tag{24}$$

If now **A** is partitioned into, say, four sub-matrices

$$\mathbf{A} = \begin{pmatrix} \boldsymbol{\alpha}_{11} & \boldsymbol{\alpha}_{12} \\ \boldsymbol{\alpha}_{21} & \boldsymbol{\alpha}_{22} \end{pmatrix} \tag{25}$$

and **B** into

$$\mathbf{B} = \begin{pmatrix} \boldsymbol{\beta}_{11} & \boldsymbol{\beta}_{12} \\ \boldsymbol{\beta}_{21} & \boldsymbol{\beta}_{22} \end{pmatrix} \tag{26}$$

then the product $\mathbf{AB} = \mathbf{C}$ may be written as

$$\mathbf{C} = \begin{pmatrix} \boldsymbol{\gamma}_{11} & \boldsymbol{\gamma}_{12} \\ \boldsymbol{\gamma}_{21} & \boldsymbol{\gamma}_{22} \end{pmatrix} \tag{27}$$

where

$$\boldsymbol{\gamma}_{ik} = \sum_{s=1}^{r} \boldsymbol{\alpha}_{is} \boldsymbol{\beta}_{sk}. \tag{28}$$

(r depending on the partitioning pattern)

provided the sub-matrices $\boldsymbol{\alpha}_{is}$ and $\boldsymbol{\beta}_{sk}$ are conformable to the product $\boldsymbol{\alpha}_{is}\boldsymbol{\beta}_{sk}$. This will always be so provided the partitioning of **A** and **B** is such that the columns of **A** are partitioned in the same way as the rows of **B**. However, the rows of **A** and the columns of **B** may be partitioned in any way whatsoever.

Example 1. To evaluate **AB** given

$$\mathbf{A} = \begin{pmatrix} 3 & 1 & 2 \\ 1 & 2 & 3 \\ 0 & 1 & 4 \end{pmatrix}, \quad \mathbf{B} = \begin{pmatrix} 1 & 3 & 0 \\ 2 & 1 & 2 \\ 3 & 2 & 1 \end{pmatrix}. \tag{29}$$

Partitioning so that

$$\mathbf{A} = \begin{pmatrix} \alpha_{11} & \alpha_{12} \\ \alpha_{21} & \alpha_{22} \end{pmatrix} = \left(\begin{array}{cc|c} 3 & 1 & 2 \\ 1 & 2 & 3 \\ \hline 0 & 1 & 4 \end{array}\right) \tag{30}$$

and

$$\mathbf{B} = \begin{pmatrix} \beta_{11} \\ \beta_{21} \end{pmatrix} = \left(\begin{array}{ccc} 1 & 3 & 0 \\ 2 & 1 & 2 \\ \hline 3 & 2 & 1 \end{array}\right) \tag{31}$$

(i.e. partitioning the columns of **A** in the same way as the rows of **B**), we have using (28)

$$\mathbf{AB} = \begin{pmatrix} \alpha_{11}\beta_{11} + \alpha_{12}\beta_{21} \\ \alpha_{21}\beta_{11} + \alpha_{22}\beta_{21} \end{pmatrix} \tag{32}$$

$$= \left(\begin{array}{c} \begin{pmatrix} 3 & 1 \\ 1 & 2 \end{pmatrix}\begin{pmatrix} 1 & 3 & 0 \\ 2 & 1 & 2 \end{pmatrix} + \begin{pmatrix} 2 \\ 3 \end{pmatrix}(3 \quad 2 \quad 1) \\ \hline (0 \quad 1)\begin{pmatrix} 1 & 3 & 0 \\ 2 & 1 & 2 \end{pmatrix} + (4)(3 \quad 2 \quad 1) \end{array}\right) \tag{33}$$

$$= \left(\begin{array}{c} \begin{pmatrix} 5 & 10 & 2 \\ 5 & 5 & 4 \end{pmatrix} + \begin{pmatrix} 6 & 4 & 2 \\ 9 & 6 & 3 \end{pmatrix} \\ \hline (2 \quad 1 \quad 2) + (12 \quad 8 \quad 4) \end{array}\right) \tag{34}$$

$$= \begin{pmatrix} 11 & 14 & 4 \\ 14 & 11 & 7 \\ 14 & 9 & 6 \end{pmatrix}. \tag{35}$$

The same result could have been obtained by partitioning **A** and as

$$\mathbf{A} = \left(\begin{array}{c|cc} 3 & 1 & 2 \\ \hline 1 & 2 & 3 \\ 0 & 1 & 4 \end{array}\right), \quad \mathbf{B} = \left(\begin{array}{cc|c} 1 & 3 & 0 \\ \hline 2 & 1 & 2 \\ 3 & 2 & 1 \end{array}\right), \tag{36}$$

where again the columns of **A** are partitioned in the same way as the rows of **B**.

2.3 Some special types of matrices

(a) *Row matrix*

A set of n quantities arranged in a row is a matrix of order $(1 \times n)$. Such a matrix is usually called a row matrix or row vector and is denoted by

$$[\mathbf{A}] = (a_1 \quad a_2 \quad a_3 \quad \dots \quad a_n). \tag{37}$$

(b) *Column matrix*

A set of m quantities arranged in a column is a matrix of order $(m \times 1)$. Such a matrix is called a column matrix or column vector and is denoted by

$$\{\mathbf{A}\} = \begin{pmatrix} a_1 \\ a_2 \\ a_3 \\ \cdot \\ \cdot \\ \cdot \\ a_m \end{pmatrix}. \tag{38}$$

(c) *Zero (or null) matrix*

A matrix of order $(m \times n)$ with all its elements equal to zero is called the zero (or null) matrix of order $(m \times n)$. For example, the matrix $\mathbf{0}$, where

$$\mathbf{0} = \begin{pmatrix} 0 & 0 & 0 \\ 0 & 0 & 0 \end{pmatrix} \tag{39}$$

is the (2×3) zero matrix.

We note here that, if $\mathbf{A}$ and $\mathbf{B}$ are two matrices conformable to the product $\mathbf{AB}$ and such that $\mathbf{AB} = \mathbf{0}$ where $\mathbf{0}$ is the zero matrix, this does not necessarily imply that either $\mathbf{A} = \mathbf{0}$ or $\mathbf{B} = \mathbf{0}$, or both. For if

$$\mathbf{A} = \begin{pmatrix} 2 & 1 \\ 4 & 2 \end{pmatrix}, \qquad \mathbf{B} = \begin{pmatrix} 1 & 3 \\ -2 & -6 \end{pmatrix} \tag{40}$$

then

$$\mathbf{AB} = \begin{pmatrix} 0 & 0 \\ 0 & 0 \end{pmatrix}. \tag{41}$$

Here again matrices do not follow the behaviour of numbers.

The zero matrix **0** has the following obvious properties:

$$\left.\begin{aligned} \mathbf{A}-\mathbf{A} &= \mathbf{0}, \\ \mathbf{A}+\mathbf{0} &= \mathbf{0}+\mathbf{A} = \mathbf{A}, \\ \mathbf{A0} &= \mathbf{0}, \\ \mathbf{0A} &= \mathbf{0}, \end{aligned}\right\} \tag{42}$$

provided **A** and **0** are conformable to the various sums and products.

(d) *Square matrices*

A matrix with the same number of rows as columns is said to be square, and to be of order n if n is the number of rows. For example, the $(n \times n)$ matrix

$$\mathbf{A} = \begin{pmatrix} a_{11} & a_{12} & \cdot & \cdot & \cdot & a_{1n} \\ a_{21} & a_{22} & \cdot & \cdot & \cdot & a_{2n} \\ \cdot & \cdot & & & & \cdot \\ \cdot & \cdot & & & & \cdot \\ \cdot & \cdot & & & & \cdot \\ a_{n1} & a_{n2} & \cdot & \cdot & \cdot & a_{nn} \end{pmatrix} \tag{43}$$

is a square matrix of order n. The diagonal containing the elements $a_{11}, a_{22}, a_{33}, \ldots, a_{nn}$ is called the leading diagonal, and the sum of these diagonal elements

$$\sum_{i=1}^{n} a_{ii}$$

is called the *trace* or *spur* of the matrix. This sum is usually denoted by *Tr* (or *Sp*). For example, if

$$\mathbf{A} = \begin{pmatrix} 1 & 0 & 1 \\ 2 & 1 & 3 \\ 3 & 4 & 5 \end{pmatrix} \tag{44}$$

then

$$Tr\,\mathbf{A} = 1+1+5 = 7. \tag{45}$$

(e) *Diagonal matrix*

A square matrix with zero elements everywhere except in the leading diagonal is called a diagonal matrix. In other words, if a_{ik} are to be the elements of a diagonal matrix we must have $a_{ik} = 0$ for $i \neq k$.

This is best described by introducing the Kronecker delta δ_{ik} which is defined by

$$\delta_{ik} = \begin{cases} 1 \text{ for } i = k, \\ 0 \text{ for } i \neq k. \end{cases} \tag{46}$$

The elements of the nth order diagonal matrix **A**, where

$$\mathbf{A} = \begin{pmatrix} \lambda_1 & 0 & 0 & \cdots & 0 \\ 0 & \lambda_2 & 0 & & \cdot \\ 0 & 0 & \lambda_3 & & \cdot \\ \cdot & & & & \cdot \\ \cdot & & & & \cdot \\ \cdot & & & & \cdot \\ 0 & \cdot & \cdot & \cdots & \lambda_n \end{pmatrix} \tag{47}$$

may now be written as

$$a_{ik} = \lambda_i \delta_{ik}, \tag{48}$$

where i and k run from 1 to n.

All diagonal matrices of the same order commute under multiplication to give another diagonal matrix. Furthermore, if **A** is the diagonal matrix of (47) and **B** is a general $(n \times n)$ matrix with elements b_{ik}, then

$$\mathbf{AB} = \begin{pmatrix} \lambda_1 & 0 & 0 & \cdots & 0 \\ 0 & \lambda_2 & 0 & \cdots & \cdot \\ 0 & 0 & & & \cdot \\ \cdot & \cdot & & & \cdot \\ \cdot & \cdot & & & \cdot \\ \cdot & \cdot & & & \cdot \\ 0 & 0 & \cdots & \cdots & \lambda_n \end{pmatrix} \begin{pmatrix} b_{11} & b_{12} & \cdots & b_{1n} \\ b_{21} & b_{22} & \cdots & b_{2n} \\ \cdot & \cdot & & \cdot \\ \cdot & \cdot & & \cdot \\ \cdot & \cdot & & \cdot \\ b_{n1} & b_{n2} & \cdots & b_{nn} \end{pmatrix} \tag{49}$$

$$= \begin{pmatrix} \lambda_1 b_{11} & \lambda_1 b_{12} & \cdots & \lambda_1 b_{1n} \\ \lambda_2 b_{21} & \lambda_2 b_{22} & \cdots & \lambda_2 b_{2n} \\ \cdot & \cdot & & \cdot \\ \cdot & \cdot & & \cdot \\ \cdot & \cdot & & \cdot \\ \lambda_n b_{n1} & \lambda_n b_{n2} & \cdots & \lambda_n b_{nn} \end{pmatrix} \tag{50}$$

We see that, on forming the product **AB**, λ_1 multiplies all the elements of the first row of **B**, λ_2 the elements of the second row, and so on. On the other hand

$$\mathbf{BA} = \begin{pmatrix} b_{11} & b_{12} & \cdot & \cdot & \cdot & b_{1n} \\ b_{21} & b_{22} & \cdot & \cdot & \cdot & b_{2n} \\ \cdot & \cdot & & & & \cdot \\ \cdot & \cdot & & & & \cdot \\ \cdot & \cdot & & & & \cdot \\ b_{n1} & b_{n2} & \cdot & \cdot & \cdot & b_{nn} \end{pmatrix} \begin{pmatrix} \lambda_1 & 0 & \cdot & \cdot & \cdot & 0 \\ 0 & \lambda_2 & \cdot & \cdot & \cdot & \cdot \\ 0 & 0 & & & & \cdot \\ \cdot & \cdot & & & & \cdot \\ \cdot & \cdot & & & & \cdot \\ \cdot & \cdot & & & & \cdot \\ 0 & 0 & \cdot & \cdot & \cdot & \lambda_n \end{pmatrix} \tag{51}$$

$$= \begin{pmatrix} \lambda_1 b_{11} & \lambda_2 b_{12} & \cdot & \cdot & \cdot & \lambda_n b_{1n} \\ \lambda_1 b_{21} & \lambda_2 b_{22} & \cdot & \cdot & \cdot & \lambda_n b_{2n} \\ \cdot & \cdot & & & & \cdot \\ \cdot & \cdot & & & & \cdot \\ \cdot & \cdot & & & & \cdot \\ \lambda_1 b_{n1} & \lambda_2 b_{n2} & \cdot & \cdot & \cdot & \lambda_n b_{nn} \end{pmatrix}, \tag{52}$$

which shows that, on forming the product **BA**, λ_1 multiplies all the elements of the first column of **B**, λ_2 the elements of the second column, and so on.

(f) *Unit matrix*

The unit (or identity) matrix of order n is an nth order diagonal matrix with unit elements. Such matrices are usually denoted by **I** (or sometimes by **E**). For example, the unit matrix of order 3 is

$$\mathbf{I} = \begin{pmatrix} 1 & 0 & 0 \\ 0 & 1 & 0 \\ 0 & 0 & 1 \end{pmatrix} \tag{53}$$

the elements of which are δ_{ik} $(i, k = 1, 2, 3)$. Such matrices are related to the identity mapping

$$\left.\begin{aligned} y_1 &= x_1, \\ y_2 &= \quad x_2, \\ y_3 &= \qquad x_3, \\ &\cdot \\ &\cdot \\ &\cdot \\ y_n &= \qquad\qquad x_n, \end{aligned}\right\} \tag{54}$$

which in matrix form reads

$$\mathbf{Y} = \mathbf{IX}. \tag{55}$$

where

$$\mathbf{Y} = \begin{pmatrix} y_1 \\ y_2 \\ \cdot \\ \cdot \\ \cdot \\ y_n \end{pmatrix}, \quad \mathbf{I} = \begin{pmatrix} 1 & 0 & 0 & \cdot & \cdot & \cdot & 0 \\ 0 & 1 & 0 & & & & \cdot \\ 0 & 0 & 1 & & & & \cdot \\ \cdot & & & & & & \cdot \\ \cdot & & & & & & \cdot \\ \cdot & & & & & & \cdot \\ 0 & \cdot & \cdot & \cdot & \cdot & \cdot & 1 \end{pmatrix} \quad \text{and } \mathbf{X} = \begin{pmatrix} x_1 \\ x_2 \\ \cdot \\ \cdot \\ \cdot \\ x_n \end{pmatrix}. \tag{56}$$

In general, if **A** is an arbitrary square matrix of order n and **I** is the unit matrix of the same order then $\mathbf{IA} = \mathbf{AI} = \mathbf{A}$. To prove this result we write $\mathbf{B} = \mathbf{IA}$. If the elements of **B** are b_{ik} then

$$b_{ik} = \sum_{s=1}^{n} \delta_{is} a_{sk} = a_{ik}. \tag{57}$$

Hence $\mathbf{IA} = \mathbf{A}$. Similarly we may prove that $\mathbf{AI} = \mathbf{A}$.

By taking $\mathbf{A} = \mathbf{I}$ in the results, we find

$$\mathbf{I}^2 = \mathbf{I} \tag{58}$$

and consequently

$$\mathbf{I}^k = \mathbf{I}^{k-1} = \ldots = \mathbf{I}^2 = \mathbf{I}, \tag{59}$$

where k is any positive integer.

If **A** is not square then $\mathbf{IA} \neq \mathbf{AI}$ since one or other of these products will not be defined. However, provided the product is defined we can always multiply an $(m \times n)$ matrix by a unit matrix without changing its form. For example,

$$\begin{pmatrix} 1 & 0 & 0 \\ 0 & 1 & 0 \\ 0 & 0 & 1 \end{pmatrix} \begin{pmatrix} 1 & 4 \\ 2 & 5 \\ 3 & 6 \end{pmatrix} = \begin{pmatrix} 1 & 4 \\ 2 & 5 \\ 3 & 6 \end{pmatrix}. \tag{60}$$

(g) *Idempotent and nilpotent matrices*

A square matrix **A** which satisfies the relation

$$\mathbf{A}^2 = \mathbf{A} \tag{61}$$

is called idempotent. Such matrices arise, for example, from the two relations

$$\mathbf{AB} = \mathbf{A} \quad \text{and} \quad \mathbf{BA} = \mathbf{B}. \tag{62}$$

For then

$$\mathbf{ABA} \begin{cases} = (\mathbf{AB})\mathbf{A} = \mathbf{A}^2, \\ = \mathbf{A}(\mathbf{BA}) = \mathbf{AB} = \mathbf{A}, \end{cases} \tag{63}$$

showing that $\mathbf{A}^2 = \mathbf{A}$.

A square matrix $\mathbf{A}$ which satisfies the relation

$$\mathbf{A}^k = \mathbf{0}, \tag{64}$$

where k is any positive integer, is said to be nilpotent of order k. For example,

$$\mathbf{A} = \begin{pmatrix} 2 & -1 \\ 4 & -2 \end{pmatrix} \tag{65}$$

is nilpotent of order 2 since $\mathbf{A}^2 = \mathbf{0}$.

(h) *The transposed matrix*

If $\mathbf{A}$ is a $(m \times n)$ matrix its transpose $\tilde{\mathbf{A}}$ (sometimes denoted by $\mathbf{A}'$ or $\mathbf{A}^{\mathrm{T}}$) is defined as the $(n \times m)$ matrix obtained by interchanging the rows and columns of $\mathbf{A}$. Consequently the ith row of $\mathbf{A}$ becomes the ith column of $\tilde{\mathbf{A}}$. For example, if

$$\mathbf{A} = \begin{pmatrix} 1 & 4 \\ 2 & 5 \\ 3 & 6 \end{pmatrix} \quad \text{then} \quad \tilde{\mathbf{A}} = \begin{pmatrix} 1 & 2 & 3 \\ 4 & 5 & 6 \end{pmatrix}. \tag{66}$$

Clearly the transpose of a column vector, say

$$\{\mathbf{A}\} = \begin{pmatrix} a_1 \\ a_2 \\ a_3 \end{pmatrix},$$

is a row vector

$$\widetilde{\{\mathbf{A}\}} = (a_1 \quad a_2 \quad a_3) = [\mathbf{A}]. \tag{67}$$

Similarly the transpose of the row vector $[\mathbf{A}]$ is the column vector $\{\mathbf{A}\}$. It follows that

$$\widetilde{\{\mathbf{A}\}}\{\mathbf{A}\} = [\mathbf{A}]\widetilde{[\mathbf{A}]} = a_1^2 + a_2^2 + a_3^2. \tag{68}$$

In general we see that if $\mathbf{A}$ is matrix of order $(m \times n)$ then $\tilde{\mathbf{A}}$ is of order $(n \times m)$ and hence $\mathbf{A}$ and $\tilde{\mathbf{A}}$ are conformable to both products $\mathbf{A}\tilde{\mathbf{A}}$ and $\tilde{\mathbf{A}}\mathbf{A}$. (N.B. both products exist but are of different orders unless $\mathbf{A}$ is square.)

We now show that if $\mathbf{A}$ and $\mathbf{B}$ are two matrices conformable to the product $\mathbf{AB} = \mathbf{C}$, then

$$\tilde{\mathbf{C}} = (\widetilde{\mathbf{AB}}) = \tilde{\mathbf{B}}\tilde{\mathbf{A}}. \tag{69}$$

To prove this, suppose $\mathbf{A}$ is of order $(m \times p)$ with elements a_{ik} and $\mathbf{B}$ is of order $(p \times n)$ with elements b_{ik}. Then $\mathbf{C}$ is a matrix of order

$(m \times n)$ with elements c_{ik} given by

$$c_{ik} = \sum_{s=1}^{p} a_{is} b_{sk}. \tag{70}$$

Consequently

$$\tilde{c}_{ik} = c_{ki} = \sum_{s=1}^{p} a_{ks} b_{si} = \sum_{s=1}^{p} \tilde{b}_{is} \tilde{a}_{sk} \tag{71}$$

and therefore

$$\tilde{\mathbf{C}} = \tilde{\mathbf{B}}\tilde{\mathbf{A}} = (\widetilde{\mathbf{AB}}). \tag{72}$$

Similarly we may show that if $\mathbf{A}_1, \mathbf{A}_2, \ldots, \mathbf{A}_n$ are n matrices conformable to the product $\mathbf{A}_1 \mathbf{A}_2 \ldots \mathbf{A}_n$, then

$$(\widetilde{\mathbf{A}_1 \mathbf{A}_2 \mathbf{A}_3 \ldots \mathbf{A}_n}) = \tilde{\mathbf{A}}_n \ldots \tilde{\mathbf{A}}_3 \tilde{\mathbf{A}}_2 \tilde{\mathbf{A}}_1. \tag{73}$$

In other words, in taking the transpose of matrix products the order of the matrices forming the product must be reversed. For example, if

$$\mathbf{A} = \begin{pmatrix} 1 & 2 \\ 3 & 4 \end{pmatrix}, \qquad \mathbf{B} = \begin{pmatrix} 2 \\ 1 \end{pmatrix} \tag{74}$$

then

$$\tilde{\mathbf{A}} = \begin{pmatrix} 1 & 3 \\ 2 & 4 \end{pmatrix}, \qquad \tilde{\mathbf{B}} = (2 \quad 1). \tag{75}$$

Consequently

$$\mathbf{AB} = \begin{pmatrix} 4 \\ 10 \end{pmatrix}, \qquad \tilde{\mathbf{B}}\tilde{\mathbf{A}} = (4 \quad 10), \tag{76}$$

and hence $(\widetilde{\mathbf{AB}}) = \tilde{\mathbf{B}}\tilde{\mathbf{A}}$.

(i) *Complex conjugate of a matrix; real and imaginary matrices*

If $\mathbf{A}$ is a matrix of order $(m \times n)$ with complex elements a_{ik} then the complex conjugate $\mathbf{A}^*$ (sometimes $\bar{\mathbf{A}}$) of $\mathbf{A}$ is found by taking the complex conjugates of all the elements. For example, if

$$\mathbf{A} = \begin{pmatrix} 1+i & 2 & -i \\ 3 & 1-i & 2+i \end{pmatrix} \quad \text{then} \quad \mathbf{A}^* = \begin{pmatrix} 1-i & 2 & i \\ 3 & 1+i & 2-i \end{pmatrix}. \tag{77}$$

It is easily seen that

$$\begin{aligned} (\mathbf{A}^*)^* &= \mathbf{A}, \\ (\lambda\mathbf{A})^* &= \lambda^*\mathbf{A}^*, \\ (\mathbf{AB})^* &= \mathbf{A}^*\mathbf{B}^*, \end{aligned} \tag{78}$$

where λ is a complex number, and where the product $\mathbf{AB}$ is assumed to exist.

A matrix $\mathbf{A}$ which satisfies the relation

$$\mathbf{A} = \mathbf{A}^* \tag{79}$$

is called real, since (79) ensures that all its elements will be real numbers. Likewise a matrix $\mathbf{A}$ is called imaginary if it satisfies the relation

$$\mathbf{A} = -\mathbf{A}^* \tag{80}$$

since this condition ensures that all its elements will be imaginary numbers.

(j) *Symmetric and skew-symmetric matrices*

A matrix $\mathbf{A}$ is symmetric if

$$\mathbf{A} = \tilde{\mathbf{A}} \quad (\text{i.e. } a_{ik} = a_{ki} \text{ for all } i, k). \tag{81}$$

Such a matrix is necessarily square and has the leading diagonal as a line of symmetry. For example, a typical symmetric matrix is

$$\mathbf{A} = \begin{pmatrix} 1 & x & y \\ x & 3 & z \\ y & z & 4 \end{pmatrix}. \tag{82}$$

For an arbitrary square matrix of order n there are n^2 independent elements. Imposing the symmetry condition $a_{ik} = a_{ki}$, however, reduces this number to

$$\frac{n^2-n}{2} + n = \tfrac{1}{2}n(n+1). \tag{83}$$

A matrix $\mathbf{A}$ is skew-symmetric if

$$\mathbf{A} = -\tilde{\mathbf{A}} \quad (\text{i.e. if } a_{ik} = -a_{ki} \text{ for all } i, k). \tag{84}$$

Such a matrix is again necessarily square, and in virtue of the relations $a_{11} = -a_{11}$, $a_{22} = -a_{22}$, etc., all the elements of the leading diagonal are zero. For example,

$$\mathbf{A} = \begin{pmatrix} 0 & 2 & 1 \\ -2 & 0 & 3 \\ -1 & -3 & 0 \end{pmatrix} \tag{85}$$

is skew-symmetric. It is easily verified that for an nth order skew-symmetric matrix the number of independent components is

$$\tfrac{1}{2}n(n-1). \tag{86}$$

Any square matrix may be written as the sum of a symmetric matrix and a skew-symmetric matrix since

$$\mathbf{A} = \left(\frac{\mathbf{A}+\tilde{\mathbf{A}}}{2}\right) + \left(\frac{\mathbf{A}-\tilde{\mathbf{A}}}{2}\right), \tag{87}$$

the first bracket being a symmetric matrix (satisfying (81)) and the second bracket a skew-symmetric matrix. We note that the sum of the numbers in (83) and (86) gives n^2 as it should in virtue of (87).

(k) *Hermitian and skew-Hermitian matrices*

A matrix **A** is called Hermitian if

$$\mathbf{A} = (\widetilde{\mathbf{A}^*}). \tag{88}$$

Such a matrix is necessarily square. We usually denote $(\widetilde{\mathbf{A}^*})$ by $\mathbf{A}^\dagger$ (or $\mathbf{A}^{\mathrm{H}}$); so **A** is Hermitian if

$$\mathbf{A} = \mathbf{A}^\dagger. \tag{89}$$

In terms of the elements a_{ik} of **A**, (89) means $a_{ik} = a_{ki}^*$, which clearly shows that the diagonal elements of a Hermitian matrix are real. For example,

$$\mathbf{A} = \begin{pmatrix} 1 & 1+i & i \\ 1-i & 2 & 4 \\ -i & 4 & 3 \end{pmatrix} \tag{90}$$

is Hermitian.

We note here that if in (88) **A** is real the definition becomes that of a symmetric matrix (see 2.3(j)).

A matrix **A** is skew-Hermitian (or anti-Hermitian) if

$$\mathbf{A} = -(\widetilde{\mathbf{A}^*}) \tag{91}$$

which, in terms of $\mathbf{A}^\dagger$, reads

$$\mathbf{A} = -\mathbf{A}^\dagger. \tag{92}$$

In terms of elements, (92) means $a_{ik} = -a_{ki}^*$, from which it follows that the diagonal elements of a skew-Hermitian matrix are either zero or purely imaginary. For example,

$$\mathbf{A} = \begin{pmatrix} i & 1 & 1-i \\ -1 & 0 & i \\ -1-i & i & 2i \end{pmatrix} \tag{93}$$

is skew-Hermitian.

Every square matrix with complex elements may be written as the sum of a Hermitian matrix and a skew-Hermitian matrix, since

$$\mathbf{A} = \tfrac{1}{2}(\mathbf{A}+\mathbf{A}^\dagger)+\tfrac{1}{2}(\mathbf{A}-\mathbf{A}^\dagger), \tag{94}$$

the first bracket being a Hermitian matrix (satisfying (89)), and the second bracket a skew-Hermitian matrix (satisfying (92)).

Now since $(\mathbf{AB})^* = \mathbf{A}^*\mathbf{B}^*$, and $\widetilde{(\mathbf{AB})} = \tilde{\mathbf{B}}\tilde{\mathbf{A}}$ (see (69)), we have

$$\widetilde{(\mathbf{AB})^*} = \widetilde{(\mathbf{A}^*\mathbf{B}^*)} = \tilde{\mathbf{B}}^*\tilde{\mathbf{A}}^* \tag{95}$$

or rather

$$(\mathbf{AB})^\dagger = \mathbf{B}^\dagger\mathbf{A}^\dagger. \tag{96}$$

PROBLEMS 2

1. If $$\mathbf{A} = \begin{pmatrix} 1 & 2 \\ 3 & 4 \end{pmatrix} \quad \text{and} \quad \mathbf{B} = \begin{pmatrix} 2 & 1 \\ 4 & 3 \end{pmatrix},$$
evaluate $(\mathbf{A}+\mathbf{B})$, $(\mathbf{A}-\mathbf{B})$, $(\mathbf{A}-\mathbf{B})(\mathbf{A}+\mathbf{B})$ and $\mathbf{A}^2-\mathbf{B}^2$.

2. If $$\mathbf{A} = \begin{pmatrix} 2 & 1 & 2 \\ 3 & 5 & 7 \\ 1 & 0 & 1 \end{pmatrix} \quad \text{and} \quad \mathbf{B} = \begin{pmatrix} -3 & 1 & 0 \\ 6 & 2 & 1 \\ 1 & -1 & 2 \end{pmatrix}$$
evaluate $\mathbf{A}+\mathbf{B}$, $\mathbf{A}-\mathbf{B}$, $\mathbf{AB}$ and $\mathbf{BA}$.

3. If $$\mathbf{A} = \begin{pmatrix} 2 & 3 \\ 4 & -1 \end{pmatrix} \quad \text{and} \quad \mathbf{B} = \begin{pmatrix} 3 & -2 \\ 2 & 1 \end{pmatrix},$$
find $\mathbf{AB}$ and $\mathbf{BA}$.
If $$\mathbf{C} = \begin{pmatrix} 1 & 2 \\ 3 & 4 \end{pmatrix}$$
verify that $\mathbf{A}(\mathbf{BC}) = (\mathbf{AB})\mathbf{C}$, and that $(\mathbf{A}+\mathbf{B})\mathbf{C} = \mathbf{AC}+\mathbf{BC}$.

4. If $$\mathbf{A} = \begin{pmatrix} 2 & 3 & 1 \\ 0 & 1 & -1 \\ -1 & 0 & 2 \end{pmatrix}, \quad \mathbf{u} = \begin{pmatrix} 1 \\ 1 \\ 0 \end{pmatrix} \quad \text{and} \quad \mathbf{v} = \begin{pmatrix} 1 \\ 1 \\ -2 \end{pmatrix}$$
calculate $\mathbf{Au}$, $\mathbf{A}^2\mathbf{u}$, $\mathbf{Av}$, $\mathbf{A}^2\mathbf{v}$ and $\tilde{\mathbf{u}}\mathbf{A}^2\mathbf{v}$.

5. Given
$$\boldsymbol{\sigma}_1 = \begin{pmatrix} 0 & 1 \\ 1 & 0 \end{pmatrix}, \quad \boldsymbol{\sigma}_2 = \begin{pmatrix} 0 & -i \\ i & 0 \end{pmatrix}, \quad \boldsymbol{\sigma}_3 = \begin{pmatrix} 1 & 0 \\ 0 & -1 \end{pmatrix},$$

show that $\sigma_1\sigma_2 = i\sigma_3$, $\sigma_2\sigma_3 = i\sigma_1$, $\sigma_3\sigma_1 = i\sigma_2$, and that $\sigma_i\sigma_k+\sigma_k\sigma_i = 2\delta_{ik}\mathbf{I}$ $(i, k = 1, 2, 3)$, where $\mathbf{I}$ is the unit matrix of order 2.

6. Given

$$\gamma_1 = \begin{pmatrix} 0 & 0 & 0 & 1 \\ 0 & 0 & 1 & 0 \\ 0 & 1 & 0 & 0 \\ 1 & 0 & 0 & 0 \end{pmatrix}, \quad \gamma_2 = \begin{pmatrix} 0 & 0 & 0 & i \\ 0 & 0 & -i & 0 \\ 0 & i & 0 & 0 \\ -i & 0 & 0 & 0 \end{pmatrix},$$

$$\gamma_3 = \begin{pmatrix} 0 & 0 & 1 & 0 \\ 0 & 0 & 0 & -1 \\ 1 & 0 & 0 & 0 \\ 0 & -1 & 0 & 0 \end{pmatrix}, \quad \gamma_4 = \begin{pmatrix} 1 & 0 & 0 & 0 \\ 0 & 1 & 0 & 0 \\ 0 & 0 & -1 & 0 \\ 0 & 0 & 0 & -1 \end{pmatrix},$$

show that $\gamma_i\gamma_k+\gamma_k\gamma_i = 2\delta_{ik}\mathbf{I}$ $(i, k = 1, 2, 3, 4)$, where $\mathbf{I}$ is the unit matrix of order 4.

7. If

$$\mathbf{A} = \begin{pmatrix} 1 & 0 & 0 & 0 \\ 1 & -1 & 0 & 0 \\ 1 & -2 & 1 & 0 \\ 1 & -3 & 3 & -1 \end{pmatrix}$$

prove that $\mathbf{A}^2 = \mathbf{I}$. Prove also that if $\mathbf{P} = \mathbf{AM}_1\mathbf{A}$ and $\mathbf{Q} = \mathbf{AM}_2\mathbf{A}$, where $\mathbf{M}_1$ and $\mathbf{M}_2$ are arbitrary diagonal matrices of order 4, then $\mathbf{PQ} = \mathbf{QP}$.

8. Find the symmetric and skew-symmetric parts of the matrix

$$\mathbf{A} = \begin{pmatrix} 1 & \frac{3}{2} & -5 \\ \frac{1}{2} & 0 & \frac{3}{4} \\ -1 & \frac{1}{4} & 2 \end{pmatrix}.$$

9. Verify that the matrix

$$\begin{pmatrix} 1 & 3-i \\ 3+i & 2 \end{pmatrix}$$

is Hermitian.

10. Prove that if $\mathbf{A}$ is skew-Hermitian, then $\pm i\mathbf{A}$ is Hermitian.

11. Determine the nature (symmetric, skew-symmetric, Hermitian or skew-Hermitian) of the following matrices:

$$\begin{pmatrix} 0 & i \\ i & 0 \end{pmatrix}, \begin{pmatrix} 1 & 2 \\ 2 & 3 \end{pmatrix}, \begin{pmatrix} 2 & 1-i \\ 1+i & 5 \end{pmatrix}, \begin{pmatrix} 0 & 5 \\ -5 & 0 \end{pmatrix}, \begin{pmatrix} i & 1-i \\ -1-i & 0 \end{pmatrix}.$$

12. Show that if $\mathbf{A}$ is Hermitian then $\mathbf{A} = \mathbf{S}+i\mathbf{T}$, where $\mathbf{S}$ and $\mathbf{T}$ are real symmetric and skew-symmetric matrices respectively.

13. Prove that $\mathbf{A}^\dagger\mathbf{A}$ and $\mathbf{A}\mathbf{A}^\dagger$ are both Hermitian.

14. Show that both the matrices

$$\begin{pmatrix} 1 & -1 & 1 \\ 1 & -1 & 1 \\ 1 & -1 & 1 \end{pmatrix} \quad \text{and} \quad \begin{pmatrix} 1 & -2 & 1 \\ -1 & 2 & -1 \\ -2 & 4 & -2 \end{pmatrix}$$

are idempotent.

CHAPTER 3

The Inverse and Related Matrices

3.1 Introduction

In Chapter 1, 1.5 we discussed briefly the idea of a linear one-to-one transformation, illustrating it by the particular case of a two-dimensional rotation of Cartesian axes. Consider now the linear transformation

$$\left.\begin{array}{l} y_1 = a_{11}x_1 + a_{12}x_2 + \quad . \quad . \quad + a_{1n}x_n, \\ y_2 = a_{21}x_1 + a_{22}x_2 + \quad . \quad . \quad + a_{2n}x_n, \\ \quad . \qquad . \qquad . \qquad\qquad\qquad . \\ \quad . \qquad . \qquad . \qquad\qquad\qquad . \\ \quad . \qquad . \qquad . \qquad\qquad\qquad . \\ y_n = a_{n1}x_1 + a_{n2}x_2 + \quad . \quad . \quad + a_{nn}x_n, \end{array}\right\} \tag{1}$$

which may be written in matrix form as

$$\mathbf{Y} = \mathbf{AX}, \tag{2}$$

where

$$\mathbf{Y} = \begin{pmatrix} y_1 \\ y_2 \\ . \\ . \\ . \\ y_n \end{pmatrix}, \quad \mathbf{X} = \begin{pmatrix} x_1 \\ x_2 \\ . \\ . \\ . \\ x_n \end{pmatrix} \quad \text{and} \quad \mathbf{A} = \begin{pmatrix} a_{11} & a_{12} & . & . & a_{1n} \\ a_{21} & a_{22} & . & . & a_{2n} \\ . & . & & & . \\ . & . & & & . \\ . & . & & & . \\ a_{n1} & a_{n2} & . & . & a_{nn} \end{pmatrix}. \tag{3}$$

We now wish to find the inverse transformation

$$\left.\begin{array}{l} x_1 = b_{11}y_1 + b_{12}y_2 + \quad . \quad . \quad + b_{1n}y_n, \\ x_2 = b_{21}y_1 + b_{22}y_2 + \quad . \quad . \quad + b_{2n}y_n, \\ \quad . \qquad . \qquad . \qquad\qquad\qquad . \\ \quad . \qquad . \qquad . \qquad\qquad\qquad . \\ \quad . \qquad . \qquad . \qquad\qquad\qquad . \\ x_n = b_{n1}y_1 + b_{n2}y_2 + \quad . \quad . \quad + b_{nn}y_n, \end{array}\right\} \tag{4}$$

(assuming that it exists) which expresses the x_i explicitly in terms of the y_i. In matrix form (4) reads

$$\mathbf{X} = \mathbf{BY}, \tag{5}$$

where **X** and **Y** are given by (3) and where

$$\mathbf{B} = \begin{pmatrix} b_{11} & b_{12} & \cdot & \cdot & b_{1n} \\ b_{21} & b_{22} & \cdot & \cdot & b_{2n} \\ \cdot & & & & \cdot \\ \cdot & & & & \cdot \\ \cdot & & & & \cdot \\ b_{n1} & \cdot & \cdot & \cdot & b_{nn} \end{pmatrix}. \tag{6}$$

From (2) and (5), it follows that

$$\mathbf{Y} = \mathbf{ABY} \quad \text{and} \quad \mathbf{X} = \mathbf{BAX} \tag{7}$$

which in turn give

$$\mathbf{AB} = \mathbf{BA} = \mathbf{I}, \tag{8}$$

where **I** is the unit matrix of order n.

B is called the inverse matrix of **A** and is denoted by $\mathbf{A}^{-1}$. Equation (8) now becomes

$$\mathbf{AA}^{-1} = \mathbf{A}^{-1}\mathbf{A} = \mathbf{I}, \tag{9}$$

which is to be compared with equation (8), Chapter 1, where essentially the same result was derived for one-to-one mappings in general. We note that **A** and $\mathbf{A}^{-1}$ necessarily commute under multiplication. What is now required is a method of calculating $\mathbf{A}^{-1}$ given the matrix **A**. To do this we first need to discuss the adjoint of a square matrix.

3.2 The adjoint matrix

If **A** is a square matrix of order n its adjoint – denoted by adj **A** – is defined as the transposed matrix of its cofactors. Suppose A_{ik} is the cofactor of the element a_{ik} in **A** (i.e. $(-1)^{i+k}$ times the value of the determinant formed by deleting the row and column in which a_{ik} occurs). Then the matrix of cofactors is the square matrix (of the same order as **A**)

$$\begin{pmatrix} A_{11} & A_{12} & \cdot & \cdot & A_{1n} \\ A_{21} & A_{22} & \cdot & \cdot & A_{2n} \\ \cdot & & & & \cdot \\ \cdot & & & & \cdot \\ \cdot & & & & \cdot \\ A_{n1} & \cdot & \cdot & \cdot & A_{nn} \end{pmatrix}. \tag{10}$$

Consequently

$$\operatorname{adj} \mathbf{A} = \begin{pmatrix} A_{11} & A_{21} & \cdots & A_{n1} \\ A_{12} & A_{22} & \cdots & A_{n2} \\ \cdot & & & \cdot \\ \cdot & & & \cdot \\ \cdot & & & \cdot \\ A_{1n} & \cdots & \cdots & A_{nn} \end{pmatrix}. \tag{11}$$

Example 1. If

$$\mathbf{A} = \begin{pmatrix} 1 & 2 & 3 \\ 1 & 3 & 5 \\ 1 & 5 & 12 \end{pmatrix} \tag{12}$$

then the cofactor of a_{11} is $A_{11} = (-1)^{1+1}(3.12-5.5) = 11$, the cofactor of a_{12} is $A_{12} = (-1)^{1+2}(12.1-5.1) = -7$, and so on. Proceeding in this way we find

$$\operatorname{adj} \mathbf{A} = \begin{pmatrix} 11 & -9 & 1 \\ -7 & 9 & -2 \\ 2 & -3 & 1 \end{pmatrix}. \tag{13}$$

Now, returning to (11) and using the expansion property of determinants

$$\sum_{s=1}^{n} a_{is} A_{ks} = |\mathbf{A}|\, \delta_{ik}, \tag{14}$$

we find

$$\mathbf{A}(\operatorname{adj} \mathbf{A}) = \begin{pmatrix} a_{11} & a_{12} & \cdots & a_{1n} \\ a_{21} & a_{22} & \cdots & a_{2n} \\ \cdot & & & \cdot \\ \cdot & & & \cdot \\ \cdot & & & \cdot \\ a_{n1} & \cdots & \cdots & a_{nn} \end{pmatrix} \begin{pmatrix} A_{11} & A_{21} & \cdots & A_{n1} \\ A_{12} & A_{22} & \cdots & A_{n2} \\ \cdot & & & \cdot \\ \cdot & & & \cdot \\ \cdot & & & \cdot \\ A_{1n} & \cdots & \cdots & A_{nn} \end{pmatrix} \tag{15}$$

$$= \begin{pmatrix} |\mathbf{A}| & 0 & 0 & \cdots & 0 \\ 0 & |\mathbf{A}| & 0 & \cdots & 0 \\ 0 & 0 & |\mathbf{A}| & & \cdot \\ \cdot & & & & \cdot \\ \cdot & & & & \cdot \\ \cdot & & & & \cdot \\ 0 & \cdots & \cdots & \cdots & |\mathbf{A}| \end{pmatrix}, \tag{16}$$

which may be written more compactly as

$$\mathbf{A}(\text{adj } \mathbf{A}) = |\mathbf{A}|\,\mathbf{I}, \tag{17}$$

where $\mathbf{I}$ is the unit matrix of order n.

Likewise, using the result

$$\sum_{s=1}^{n} A_{sk}\, a_{si} = |\mathbf{A}|\, \delta_{ik}, \tag{18}$$

we may easily prove that

$$(\text{adj } \mathbf{A})\mathbf{A} = |\mathbf{A}|\,\mathbf{I}. \tag{19}$$

Consequently

$$\mathbf{A}(\text{adj } \mathbf{A}) = (\text{adj } \mathbf{A})\mathbf{A} = |\mathbf{A}|\,\mathbf{I}, \tag{20}$$

which shows that $\mathbf{A}$ and its adjoint matrix commute under multiplication.

Some further properties of the adjoint matrix may be derived from (20). For example, taking determinants (and remembering that the determinant of a product is the product of the determinants†) we have

$$|\mathbf{A}|\,|\text{adj } \mathbf{A}| = |\mathbf{A}|^{n} \tag{21}$$

or

$$|\text{adj } \mathbf{A}| = |\mathbf{A}|^{n-1}, \tag{22}$$

provided $|\mathbf{A}| \neq 0$.

3.3 The inverse matrix

From (20) it is clear that the matrix $\dfrac{\text{adj } \mathbf{A}}{|\mathbf{A}|}$ behaves in the way required of an inverse of $\mathbf{A}^{-1}$ since

$$\left(\frac{\text{adj } \mathbf{A}}{|\mathbf{A}|}\right)\mathbf{A} = \mathbf{A}\left(\frac{\text{adj } \mathbf{A}}{|\mathbf{A}|}\right) = \mathbf{I} \tag{23}$$

(compare with (9)). Consequently we define

$$\mathbf{A}^{-1} = \frac{\text{adj } \mathbf{A}}{|\mathbf{A}|}. \tag{24}$$

It is necessary, however, to show that this inverse is unique in that there is no other matrix with the desired properties. To prove this

† There are many ways of proving this result. It is sufficient here for the reader to establish the result for the determinants of two (2×2) matrices and to see that the method can be readily extended to higher order determinants.

suppose that $\mathbf{X}$ is any matrix such that $\mathbf{AX} = \mathbf{I}$. Then

$$\mathbf{A}^{-1}\mathbf{AX} = \mathbf{A}^{-1}\mathbf{I} = \mathbf{A}^{-1}. \tag{25}$$

Since $\mathbf{A}^{-1}\mathbf{A} = \mathbf{I}$, (25) gives $\mathbf{X} = \mathbf{A}^{-1}$. Likewise if $\mathbf{Y}$ is a matrix such that $\mathbf{YA} = \mathbf{I}$, then

$$\mathbf{YAA}^{-1} = \mathbf{IA}^{-1} = \mathbf{A}^{-1} \tag{26}$$

and, since $\mathbf{AA}^{-1} = \mathbf{I}$, $\mathbf{Y} = \mathbf{A}^{-1}$. Consequently provided the inverse exists it is unique. Clearly, from (24), the inverse $\mathbf{A}^{-1}$ exists provided $|\mathbf{A}| \neq 0$ – that is provided $\mathbf{A}$ is non-singular.

Example 2. The adjoint of the matrix

$$\mathbf{A} = \begin{pmatrix} 1 & 4 & 0 \\ -1 & 2 & 2 \\ 0 & 0 & 2 \end{pmatrix} \tag{27}$$

is the matrix

$$\begin{pmatrix} 4 & -8 & 8 \\ 2 & 2 & -2 \\ 0 & 0 & 6 \end{pmatrix}. \tag{28}$$

Furthermore $|\mathbf{A}| = 12$. Hence

$$\mathbf{A}^{-1} = \frac{\operatorname{adj} \mathbf{A}}{|\mathbf{A}|} = \begin{pmatrix} \frac{1}{3} & -\frac{2}{3} & \frac{2}{3} \\ \frac{1}{6} & \frac{1}{6} & -\frac{1}{6} \\ 0 & 0 & \frac{1}{2} \end{pmatrix}. \tag{29}$$

It may easily be verified that $\mathbf{AA}^{-1} = \mathbf{A}^{-1}\mathbf{A} = \mathbf{I}$.

3.4 Some properties of the inverse matrix

Suppose $\mathbf{A}$ and $\mathbf{B}$ are two square non-singular matrices of the same order. Then since $|\mathbf{A}|$ and $|\mathbf{B}|$ are both non-zero so also is $|\mathbf{AB}|$. Consequently $\mathbf{AB}$ has an inverse $(\mathbf{AB})^{-1}$ such that

$$(\mathbf{AB})(\mathbf{AB})^{-1} = \mathbf{I}. \tag{30}$$

Hence multiplying (30) throughout on the left by $\mathbf{B}^{-1}\mathbf{A}^{-1}$ we have

$$\mathbf{B}^{-1}\mathbf{A}^{-1}\mathbf{AB}(\mathbf{AB})^{-1} = \mathbf{B}^{-1}\mathbf{A}^{-1}\mathbf{I}, \tag{31}$$

which gives (since $\mathbf{A}^{-1}\mathbf{A} = \mathbf{I}$, $\mathbf{B}^{-1}\mathbf{B} = \mathbf{I}$)

$$(\mathbf{AB})^{-1} = \mathbf{B}^{-1}\mathbf{A}^{-1}. \tag{32}$$

The same result is obtained by taking $(\mathbf{AB})^{-1}\mathbf{AB} = \mathbf{I}$ and multiplying throughout on the right by $\mathbf{B}^{-1}\mathbf{A}^{-1}$. Equation (32) shows that the inverse of a product is obtained by taking the product of the inverses in reverse order. This result was, in fact, deduced in Chapter 1, 1.4 using the idea of general mappings, and may be extended to n non-singular matrices $\mathbf{A}_1, \mathbf{A}_2, \mathbf{A}_3, \ldots, \mathbf{A}_n$ of the same order to give

$$(\mathbf{A}_1\mathbf{A}_2\ldots\mathbf{A}_{n-1}\mathbf{A}_n)^{-1} = \mathbf{A}_n^{-1}\mathbf{A}_{n-1}^{-1}\ldots\mathbf{A}_2^{-1}\mathbf{A}_1^{-1}. \tag{33}$$

Another result which is easily proved is that if $\mathbf{A}$ is a non-singular matrix then

$$(\tilde{\mathbf{A}})^{-1} = \widetilde{(\mathbf{A}^{-1})}. \tag{34}$$

For, since $\mathbf{A}\mathbf{A}^{-1} = \mathbf{A}^{-1}\mathbf{A} = \mathbf{I}$, we have

$$\widetilde{(\mathbf{A}\mathbf{A}^{-1})} = \widetilde{(\mathbf{A}^{-1})}\tilde{\mathbf{A}} = \tilde{\mathbf{I}} = \mathbf{I} \tag{35}$$

and

$$\widetilde{(\mathbf{A}^{-1}\mathbf{A})} = \tilde{\mathbf{A}}\widetilde{(\mathbf{A}^{-1})} = \tilde{\mathbf{I}} = \mathbf{I}. \tag{36}$$

Consequently

$$\tilde{\mathbf{A}}\widetilde{(\mathbf{A}^{-1})} = \widetilde{(\mathbf{A}^{-1})}\tilde{\mathbf{A}} = \mathbf{I}. \tag{37}$$

However

$$\tilde{\mathbf{A}}(\tilde{\mathbf{A}})^{-1} = (\tilde{\mathbf{A}})^{-1}\tilde{\mathbf{A}} = \mathbf{I}, \tag{38}$$

so

$$(\tilde{\mathbf{A}})^{-1} = \widetilde{(\mathbf{A}^{-1})}. \tag{39}$$

Finally we may now show that if $\mathbf{A}$ is non-singular *and symmetric* then so also is $\mathbf{A}^{-1}$. For since

$$\mathbf{A}^{-1}\mathbf{A} = \mathbf{I} = \tilde{\mathbf{I}} = \widetilde{(\mathbf{A}\mathbf{A}^{-1})} = \widetilde{(\mathbf{A}^{-1})}\tilde{\mathbf{A}}, \tag{40}$$

it follows, using the symmetry of $\mathbf{A}$ expressed by the relation $\mathbf{A} = \tilde{\mathbf{A}}$, that

$$\mathbf{A}^{-1} = \widetilde{(\mathbf{A}^{-1})}. \tag{41}$$

Consequently $\mathbf{A}^{-1}$ is symmetric.

3.5 Evaluation of the inverse matrix by partitioning

Suppose $\mathbf{A}$ is a non-singular square matrix of order n. We now partition $\mathbf{A}$ into sub-matrices (see Chapter 2, 2.2) as

$$\mathbf{A} = \left(\begin{array}{c|c} \alpha_{11} & \alpha_{12} \\ \hline \alpha_{21} & \alpha_{22} \end{array}\right), \tag{42}$$

where $\boldsymbol{\alpha}_{11}$ is an $(s\times s)$ matrix, $\boldsymbol{\alpha}_{12}$ an $(s\times s)$ matrix, $\boldsymbol{\alpha}_{21}$ a $(s\times s)$ matrix and $\boldsymbol{\alpha}_{22}$ a $(s\times s)$ matrix, and where $2s=n$.

Let the inverse matrix $\mathbf{A}^{-1}$ be partitioned as

$$\mathbf{A}^{-1}=\left(\begin{array}{c|c}\boldsymbol{\beta}_{11} & \boldsymbol{\beta}_{12}\\ \hline \boldsymbol{\beta}_{21} & \boldsymbol{\beta}_{22}\end{array}\right), \tag{43}$$

where the partitioning is carried out in exactly the same way as the partitioning of $\mathbf{A}$ (i.e. $\boldsymbol{\beta}_{11}$ is an $(s\times s)$ matrix, etc.). Now, since $\mathbf{A}\mathbf{A}^{-1}=\mathbf{I}$, we have

$$\left(\begin{array}{c|c}\boldsymbol{\alpha}_{11} & \boldsymbol{\alpha}_{12}\\ \hline \boldsymbol{\alpha}_{21} & \boldsymbol{\alpha}_{22}\end{array}\right)\left(\begin{array}{c|c}\boldsymbol{\beta}_{11} & \boldsymbol{\beta}_{12}\\ \hline \boldsymbol{\beta}_{21} & \boldsymbol{\beta}_{22}\end{array}\right)=\left(\begin{array}{c|c}\mathbf{I}_s & 0\\ \hline 0 & \mathbf{I}_s\end{array}\right), \tag{44}$$

where $\mathbf{I}_s$ is the unit matrix of order s.

From (44) it follows that

$$\boldsymbol{\alpha}_{11}\boldsymbol{\beta}_{11}+\boldsymbol{\alpha}_{12}\boldsymbol{\beta}_{21}=\mathbf{I}_s, \tag{45}$$

$$\boldsymbol{\alpha}_{11}\boldsymbol{\beta}_{12}+\boldsymbol{\alpha}_{12}\boldsymbol{\beta}_{22}=\mathbf{0}, \tag{46}$$

$$\boldsymbol{\alpha}_{21}\boldsymbol{\beta}_{11}+\boldsymbol{\alpha}_{22}\boldsymbol{\beta}_{21}=\mathbf{0}, \tag{47}$$

$$\boldsymbol{\alpha}_{21}\boldsymbol{\beta}_{12}+\boldsymbol{\alpha}_{22}\boldsymbol{\beta}_{22}=\mathbf{I}_s. \tag{48}$$

Putting $\boldsymbol{\beta}_{22}=\mathbf{k}^{-1}$, we have from (46)

$$\boldsymbol{\beta}_{12}=-\boldsymbol{\alpha}_{11}^{-1}\boldsymbol{\alpha}_{12}\mathbf{k}^{-1} \tag{49}$$

(assuming that $\boldsymbol{\alpha}_{11}$ is non-singular), and from (48)

$$\boldsymbol{\beta}_{12}=\boldsymbol{\alpha}_{21}^{-1}-\boldsymbol{\alpha}_{21}^{-1}\boldsymbol{\alpha}_{22}\mathbf{k}^{-1} \tag{50}$$

(assuming that $\boldsymbol{\alpha}_{21}$ is non-singular also).

Comparing (49) and (50) it is easily found that

$$\mathbf{k}=\boldsymbol{\alpha}_{22}-\boldsymbol{\alpha}_{21}\boldsymbol{\alpha}_{11}^{-1}\boldsymbol{\alpha}_{12}. \tag{51}$$

Hence, using (51) and (50),

$$\boldsymbol{\beta}_{12}=-\boldsymbol{\alpha}_{11}^{-1}\boldsymbol{\alpha}_{12}\mathbf{k}^{-1}. \tag{52}$$

Now from (47)

$$\boldsymbol{\beta}_{11}=-\boldsymbol{\alpha}_{21}^{-1}\boldsymbol{\alpha}_{22}\boldsymbol{\beta}_{21} \tag{53}$$

and from (45)

$$\boldsymbol{\beta}_{11}=\boldsymbol{\alpha}_{11}^{-1}-\boldsymbol{\alpha}_{11}^{-1}\boldsymbol{\alpha}_{12}\boldsymbol{\beta}_{21}. \tag{54}$$

Consequently from (53) and (54)

$$\boldsymbol{\beta}_{21}=\mathbf{I}_s(\boldsymbol{\alpha}_{12}-\boldsymbol{\alpha}_{11}\boldsymbol{\alpha}_{21}^{-1}\boldsymbol{\alpha}_{22})^{-1} \tag{55}$$

$$=-\mathbf{k}^{-1}\boldsymbol{\alpha}_{21}\boldsymbol{\alpha}_{11}^{-1}\quad\text{(using (51))}. \tag{56}$$

Finally, using (56) and (54), we have

$$\boldsymbol{\beta}_{11} = \boldsymbol{\alpha}_{11}^{-1} + \boldsymbol{\alpha}_{11}^{-1}\boldsymbol{\alpha}_{12}\mathbf{k}^{-1}\boldsymbol{\alpha}_{21}\boldsymbol{\alpha}_{11}^{-1}. \tag{57}$$

Collecting the appropriate results together we have

$$\left.\begin{aligned} \boldsymbol{\beta}_{11} &= \boldsymbol{\alpha}_{11}^{-1} + \boldsymbol{\alpha}_{11}^{-1}\boldsymbol{\alpha}_{12}\mathbf{k}^{-1}\boldsymbol{\alpha}_{21}\boldsymbol{\alpha}_{11}^{-1}, \\ \boldsymbol{\beta}_{12} &= -\boldsymbol{\alpha}_{11}^{-1}\boldsymbol{\alpha}_{12}\mathbf{k}^{-1}, \\ \boldsymbol{\beta}_{21} &= -\mathbf{k}^{-1}\boldsymbol{\alpha}_{21}\boldsymbol{\alpha}_{11}^{-1}, \\ \boldsymbol{\beta}_{22} &= \mathbf{k}^{-1}, \end{aligned}\right\} \tag{58}$$

where $\mathbf{k}$ is defined by (51). In the calculation of these sub-matrices the inverse matrices which need be calculated are $\boldsymbol{\alpha}_{11}^{-1}$ and $\mathbf{k}^{-1}$. Identical results to (58) may be obtained by partitioning $\mathbf{A}$ according to some other pattern. However, the choice of the way in which the original matrix $\mathbf{A}$ is partitioned depends very much on its form. In general, however, the method of partitioning enables the inversion of a large matrix to be reduced to the inversion of several smaller order matrices.

Example 3. To find by partitioning the inverse of the matrix

$$\mathbf{A} = \begin{pmatrix} 1 & 2 & 1 \\ -1 & 2 & 1 \\ 2 & 5 & 3 \end{pmatrix}. \tag{59}$$

Let

$$\boldsymbol{\alpha}_{11} = \begin{pmatrix} 1 & 2 \\ -1 & 2 \end{pmatrix}, \quad \boldsymbol{\alpha}_{12} = \begin{pmatrix} 1 \\ 1 \end{pmatrix}, \tag{60}$$
$$\boldsymbol{\alpha}_{21} = (2 \quad 5), \qquad \boldsymbol{\alpha}_{22} = (3).$$

It is easily found that

$$\boldsymbol{\alpha}_{11}^{-1} = \tfrac{1}{4}\begin{pmatrix} 2 & -2 \\ 1 & 1 \end{pmatrix} \tag{61}$$

and hence that

$$k = 3 - (2 \quad 5)\tfrac{1}{4}\begin{pmatrix} 2 & -2 \\ 1 & 1 \end{pmatrix}\begin{pmatrix} 1 \\ 1 \end{pmatrix} \tag{62}$$

$$= \tfrac{1}{2}. \tag{63}$$

Consequently, using (58),

$$\boldsymbol{\beta}_{11} = \begin{pmatrix} \frac{1}{2} & -\frac{1}{2} \\ \frac{1}{4} & \frac{1}{4} \end{pmatrix} + 2\begin{pmatrix} \frac{1}{2} & -\frac{1}{2} \\ \frac{1}{4} & \frac{1}{4} \end{pmatrix}\begin{pmatrix} 1 \\ 1 \end{pmatrix}(2 \quad 5)\begin{pmatrix} \frac{1}{2} & -\frac{1}{2} \\ \frac{1}{4} & \frac{1}{4} \end{pmatrix} \tag{64}$$

$$= \begin{pmatrix} \frac{1}{2} & -\frac{1}{2} \\ \frac{5}{2} & \frac{1}{2} \end{pmatrix}, \tag{65}$$

$$\boldsymbol{\beta}_{12} = -2\begin{pmatrix} \frac{1}{2} & -\frac{1}{2} \\ \frac{1}{4} & \frac{1}{4} \end{pmatrix}\begin{pmatrix} 1 \\ 1 \end{pmatrix} = \begin{pmatrix} 0 \\ -1 \end{pmatrix}, \tag{66}$$

$$\boldsymbol{\beta}_{21} = -2(2 \quad 5)\begin{pmatrix} \frac{1}{2} & -\frac{1}{2} \\ \frac{1}{4} & \frac{1}{4} \end{pmatrix} = (-\tfrac{9}{2} \; -\tfrac{1}{2}) \tag{67}$$

and

$$\boldsymbol{\beta}_{22} = 2. \tag{68}$$

Hence the inverse of **A** is the matrix

$$\begin{pmatrix} \frac{1}{2} & -\frac{1}{2} & 0 \\ \frac{5}{2} & \frac{1}{2} & -1 \\ -\frac{9}{2} & -\frac{1}{2} & 2 \end{pmatrix}. \tag{69}$$

3.6 Orthogonal matrices and orthogonal transformations

A square matrix **A** (with real elements) is said to be orthogonal if

$$\tilde{\mathbf{A}} = \mathbf{A}^{-1}. \tag{70}$$

Since $\mathbf{AA}^{-1} = \mathbf{A}^{-1}\mathbf{A} = \mathbf{I}$, it follows that an orthogonal matrix **A** satisfies the relation

$$\mathbf{A}\tilde{\mathbf{A}} = \tilde{\mathbf{A}}\mathbf{A} = \mathbf{I}. \tag{71}$$

From (71) it is easily deduced that the columns (or column vectors) and also the rows (or row vectors) of an orthogonal matrix form an orthonormal set of vectors (i.e. mutually orthogonal and of unit length). For example, the matrix

$$\mathbf{A} = \begin{pmatrix} \frac{1}{3} & \frac{2}{3} & \frac{2}{3} \\ \frac{2}{3} & \frac{1}{3} & -\frac{2}{3} \\ -\frac{2}{3} & \frac{2}{3} & -\frac{1}{3} \end{pmatrix} \tag{72}$$

is orthogonal. Now taking the first column of elements as representing the components of a 3-vector, its length is

$$\{(\tfrac{1}{3})^2 + (\tfrac{2}{3})^2 + (-\tfrac{2}{3})^2\}^{\frac{1}{2}} = 1.$$

Likewise for the second and third columns (and also the rows). Furthermore, taking the scalar product of the first column vector with the second column vector, we have $\frac{1}{3}\cdot\frac{2}{3} + \frac{2}{3}\cdot\frac{1}{3} - \frac{2}{3} + \frac{2}{3} = 0$, showing that the first two column vectors are mutually orthogonal. Likewise for the second and third columns, and the third and first columns. Similar results hold for the rows. These results may be expressed more compactly for a general n^{th} order orthogonal matrix **A** with elements a_{ik} by the relation

$$\sum_{s=1}^{n} a_{is}a_{ks} = \delta_{ik}. \tag{73}$$

Some other results concerning orthogonal matrices may easily be proved.

(a) Suppose **A** and **B** are two n^{th} order orthogonal matrices. Then $\mathbf{A\tilde{A}} = \mathbf{\tilde{A}A} = \mathbf{I}$ and $\mathbf{B\tilde{B}} = \mathbf{\tilde{B}B} = \mathbf{I}$. Hence

$$(\mathbf{AB})(\widetilde{\mathbf{AB}}) = \mathbf{AB\tilde{B}\tilde{A}} \quad \text{(using equation (72), Chapter 2)}, \tag{74}$$

$$= \mathbf{I}. \tag{75}$$

Likewise

$$(\widetilde{\mathbf{AB}})\mathbf{AB} = \mathbf{\tilde{B}\tilde{A}AB} = \mathbf{I}, \tag{76}$$

so consequently the product of two orthogonal matrices is an orthogonal matrix.

(b) The transpose of an orthogonal matrix **A** is orthogonal. For, since $\mathbf{\tilde{A}} = \mathbf{A}^{-1}$, then

$$(\tilde{\mathbf{\tilde{A}}}) = (\widetilde{\mathbf{A}^{-1}}) = (\mathbf{\tilde{A}})^{-1} \quad \text{(using (39))}, \tag{77}$$

which shows $\mathbf{\tilde{A}}$ to be orthogonal.

(c) The inverse of an orthogonal matrix **A** is orthogonal. For, since $\mathbf{\tilde{A}} = \mathbf{A}^{-1}$, then

$$(\widetilde{\mathbf{A}^{-1}}) = (\tilde{\mathbf{\tilde{A}}}) = \mathbf{A} = (\mathbf{A}^{-1})^{-1}, \tag{78}$$

which shows $\mathbf{A}^{-1}$ to be orthogonal.

(d) The determinant of an orthogonal matrix is equal to ± 1. For, since $\mathbf{A\tilde{A}} = \mathbf{I}$, we have, taking determinants,

$$|\mathbf{A}|\,|\mathbf{\tilde{A}}| = |\mathbf{A}|^2 = 1, \tag{79}$$

so that $|\mathbf{A}| = \pm 1$.

Suppose now that the elements of the column vector

$$\mathbf{X} = \begin{pmatrix} x_1 \\ x_2 \\ \cdot \\ \cdot \\ \cdot \\ x_n \end{pmatrix} \tag{80}$$

represent the Cartesian coordinates $(x_1, x_2, \ldots, x_n)$ of a point P

in an Euclidean space of n-dimensions. Then $\tilde{\mathbf{X}}\mathbf{X} = x_1^2+x_2^2+\ldots+x_n^2$ gives the squared distance from the origin to the point P (in 3-dimensions this reduces to the well-known result $x_1^2+x_2^2+x_3^2$). Now supposing we make a coordinate transformation by means of the matrix relation $\mathbf{Y} = \mathbf{AX}$, where $\mathbf{A}$ is an nth order matrix and

$$\mathbf{Y} = \begin{pmatrix} y_1 \\ y_2 \\ \cdot \\ \cdot \\ \cdot \\ y_n \end{pmatrix}.$$

With respect to the new axes, the coordinates of the point P are now $(y_1, y_2, \ldots, y_n)$. The distance of P from the origin (which is a fixed point under the transformation, since when $\mathbf{X} = \mathbf{0}$, $\mathbf{Y} = \mathbf{0}$ also) is now $\tilde{\mathbf{Y}}\mathbf{Y} = y_1^2+y_2^2+\ldots+y_n^2$. But

$$\tilde{\mathbf{Y}}\mathbf{Y} = \widetilde{(\mathbf{AX})}\mathbf{AX} = \tilde{\mathbf{X}}\tilde{\mathbf{A}}\mathbf{AX} \quad \text{(using (72), Chapter 2)}, \tag{81}$$

so if $\tilde{\mathbf{A}}\mathbf{A} = \mathbf{I}$ then $\tilde{\mathbf{Y}}\mathbf{Y} = \tilde{\mathbf{X}}\mathbf{X}$, and distance is preserved (i.e. is an invariant quantity) under the transformation. But $\tilde{\mathbf{A}}\mathbf{A} = \mathbf{I}$ is the condition that $\mathbf{A}$ be orthogonal. Consequently the transformation

$$\mathbf{Y} = \mathbf{AX} \quad \text{with } \mathbf{A} \text{ orthogonal} \tag{82}$$

is called an orthogonal transformation, and has the important property of preserving distance. Besides leaving distance unaltered,

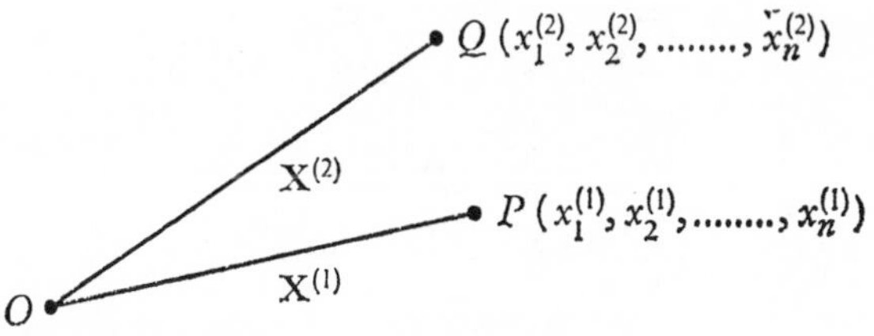

Fig. 3.1

an orthogonal transformation also leaves the angle between any two vectors unaltered. Suppose $\mathbf{X}^{(1)}$ and $\mathbf{X}^{(2)}$ are two vectors in n-dimensional space (see Fig. 3.1), and that they are represented by column matrices

$$\mathbf{X}^{(1)} = \begin{pmatrix} x_1^{(1)} \\ x_2^{(1)} \\ \cdot \\ \cdot \\ \cdot \\ x_n^{(1)} \end{pmatrix}, \qquad \mathbf{X}^{(2)} = \begin{pmatrix} x_1^{(2)} \\ x_2^{(2)} \\ \cdot \\ \cdot \\ \cdot \\ x_n^{(2)} \end{pmatrix}, \tag{83}$$

where $(x_1^{(1)}, x_2^{(1)}, \ldots, x_n^{(1)})$ and $(x_1^{(2)}, x_2^{(2)}, \ldots, x_n^{(2)})$ are coordinates of the points P and Q respectively. Then the angle θ between $\mathbf{X}^{(1)}$ and $\mathbf{X}^{(2)}$ is defined by

$$\cos\theta = \frac{x_1^{(1)}x_1^{(2)} + x_2^{(1)}x_2^{(2)} + \ldots + x_n^{(1)}x_n^{(2)}}{\{x_1^{(1)2} + x_2^{(1)2} + \ldots + x_n^{(1)2}\}^{\frac{1}{2}}\{x_1^{(2)2} + x_2^{(2)2} + \ldots + x_n^{(2)2}\}^{\frac{1}{2}}}. \tag{84}$$

in matrix form this becomes

$$\cos\theta = \frac{\tilde{\mathbf{X}}^{(1)}\mathbf{X}^{(2)}}{\{\tilde{\mathbf{X}}^{(1)}\mathbf{X}^{(1)}\}^{\frac{1}{2}}\{\tilde{\mathbf{X}}^{(2)}\mathbf{X}^{(2)}\}^{\frac{1}{2}}}. \tag{85}$$

Now let $\mathbf{X}^{(1)}$ be transformed into a new vector $\mathbf{Y}^{(1)}$ by an orthogonal transformation $\mathbf{Y}^{(1)} = \mathbf{AX}^{(1)}$, and $\mathbf{X}^{(2)}$ be transformed into a new vector $\mathbf{Y}^{(2)}$ by the same matrix so that $\mathbf{Y}^{(2)} = \mathbf{AX}^{(2)}$. Then the angle ϕ, say, between the new vectors $\mathbf{Y}^{(1)}$ and $\mathbf{Y}^{(2)}$ is

$$\cos\phi = \frac{\tilde{\mathbf{Y}}^{(1)}\mathbf{Y}^{(2)}}{\{\tilde{\mathbf{Y}}^{(1)}\mathbf{Y}^{(1)}\}^{\frac{1}{2}}\{\tilde{\mathbf{Y}}^{(2)}\mathbf{Y}^{(2)}\}^{\frac{1}{2}}} \tag{86}$$

$$= \frac{\widetilde{(\mathbf{AX}^{(1)})}\mathbf{AX}^{(2)}}{\{\widetilde{(\mathbf{AX}^{(1)})}\mathbf{AX}^{(1)}\}^{\frac{1}{2}}\{\widetilde{(\mathbf{AX}^{(2)})}\mathbf{AX}^{(2)}\}^{\frac{1}{2}}} \tag{87}$$

$$= \frac{\widetilde{\mathbf{X}^{(1)}}\tilde{\mathbf{A}}\mathbf{AX}^{(2)}}{\{\widetilde{\mathbf{X}^{(1)}}\tilde{\mathbf{A}}\mathbf{AX}^{(1)}\}^{\frac{1}{2}}\{\tilde{\mathbf{X}}^{(2)}\tilde{\mathbf{A}}\mathbf{AX}^{(2)}\}^{\frac{1}{2}}} \tag{88}$$

$$= \frac{\tilde{\mathbf{X}}^{(1)}\mathbf{X}^{(2)}}{\{\tilde{\mathbf{X}}^{(1)}\mathbf{X}^{(1)}\}^{\frac{1}{2}}\{\tilde{\mathbf{X}}^{(2)}\mathbf{X}^{(2)}\}^{\frac{1}{2}}} \quad (\text{since } \tilde{\mathbf{A}}\mathbf{A} = \mathbf{I}) \tag{89}$$

$$= \cos\theta.$$

Hence $\theta = \phi$, showing that an orthogonal transformation preserves angles between vectors.

We now consider the conditions to be placed on the general second order matrix with real elements

$$\mathbf{A} = \begin{pmatrix} a_{11} & a_{12} \\ a_{21} & a_{22} \end{pmatrix} \tag{90}$$

for it to be orthogonal. Now, since we require $\mathbf{A}\tilde{\mathbf{A}} = \mathbf{I}$, we have

$$\begin{pmatrix} a_{11} & a_{12} \\ a_{21} & a_{22} \end{pmatrix} \begin{pmatrix} a_{11} & a_{21} \\ a_{12} & a_{22} \end{pmatrix} = \begin{pmatrix} 1 & 0 \\ 0 & 1 \end{pmatrix}. \tag{91}$$

Hence

$$a_{11}^2 + a_{12}^2 = 1, \tag{92}$$

$$a_{11}a_{21} + a_{12}a_{22} = 0, \tag{93}$$

and

$$a_{21}^2 + a_{22}^2 = 1. \tag{94}$$

Writing $a_{11} = \cos\theta$, $a_{21} = \cos\phi$, (92) and (94) give $a_{12} = \sin\theta$ and $a_{22} = \sin\phi$ respectively. Equation (93) now becomes

$$\cos(\theta - \phi) = 0 \tag{95}$$

giving $\phi = \theta + \pi/2$ or $\phi = \theta + 3\pi/2$. Consequently

$$a_{22} = \pm\cos\theta, \qquad a_{21} = \mp\sin\theta. \tag{96}$$

Hence there are only two possible second-order orthogonal matrices, namely

$$\mathbf{A}_1 = \begin{pmatrix} \cos\theta & \sin\theta \\ -\sin\theta & \cos\theta \end{pmatrix} \quad \text{and} \quad \mathbf{A}_2 = \begin{pmatrix} \cos\theta & \sin\theta \\ \sin\theta & -\cos\theta \end{pmatrix}. \tag{97}$$

The first of these two matrices, namely $\mathbf{A}_1$, has been met earlier (see Chapter 1, 1.5) in connection with the rotation of Cartesian axes. The orthogonal transformation $\mathbf{Y} = \mathbf{A}_1\mathbf{X}$ corresponds therefore to a rotation about the origin. The transformation $\mathbf{Y} = \mathbf{A}_2\mathbf{X}$, however, is not just a rotation about the origin, but consists of a rotation of the axes through an angle θ together with a reversal of the sign of the second coordinate. In other words, a rotation through an angle θ, followed by a reflection in the $0y_1$ axis (see Fig. 3.2). The essential difference between these two transformations is in the values of the determinants of the matrices $\mathbf{A}_1$ and $\mathbf{A}_2$. For $|\mathbf{A}_1| = +1$, whilst $|\mathbf{A}_2| = -1$. In general, orthogonal transformations for which $|\mathbf{A}| = +1$ correspond to pure rotations about the origin, whilst orthogonal transformations with $|\mathbf{A}| = -1$ correspond to a rotation

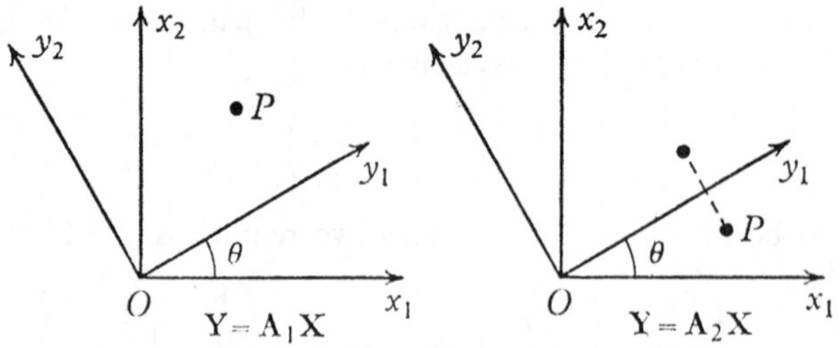

Fig. 3.2

plus a reflection in one or other of the planes defined by the axes. A rigid body is therefore unaltered in size and shape by an orthogonal transformation (since length and angle are both unaltered by the transformation), but undergoes either a pure rotation about the origin or a rotation about the origin together with a reflection in an axis plane.

3.7 Unitary matrices

A square matrix $\mathbf{A}$ is said to be unitary if

$$\mathbf{A}^\dagger = \mathbf{A}^{-1} \quad (\text{i.e. } \mathbf{A}\mathbf{A}^\dagger = \mathbf{A}^\dagger\mathbf{A} = \mathbf{I}), \tag{98}$$

where $\mathbf{A}^\dagger = (\widetilde{\mathbf{A}^*})$ is the transpose of the complex conjugate of $\mathbf{A}$ (see Chapter 2, 2.3(k)). Clearly when the elements of $\mathbf{A}$ are real (98) reduces to $\tilde{\mathbf{A}} = \mathbf{A}^{-1}$, which is the definition of an orthogonal matrix. In fact, the unitary matrix is the generalisation of the real orthogonal matrix when the matrix elements are allowed to be complex. For this reason similar results to those obtained in the last section still apply. For example, the product of two unitary matrices is an unitary matrix since, if $\mathbf{A}$ and $\mathbf{B}$ are unitary, $\mathbf{A}^\dagger = \mathbf{A}^{-1}$ and $\mathbf{B}^\dagger = \mathbf{B}^{-1}$, and hence

$$(\mathbf{AB})^\dagger = \mathbf{B}^\dagger\mathbf{A}^\dagger \quad (\text{using (96), Chapter 2}) \tag{99}$$

$$= \mathbf{B}^{-1}\mathbf{A}^{-1} = (\mathbf{AB})^{-1} \quad (\text{using (32), Chapter 3}). \tag{100}$$

As in the previous section we may also prove that the inverse and transpose of a unitary matrix are unitary matrices. The transformation $\mathbf{Y} = \mathbf{AX}$ is called unitary if the matrix $\mathbf{A}$ is unitary, and preserves lengths of vectors and angles between them when the vector components are *complex* numbers $(x_1, x_2, \ldots, x_n)$. The definition of squared distance, for example, is different now and is no longer given by $\tilde{\mathbf{X}}\mathbf{X}$ (see last section) but by

$$\widetilde{\mathbf{X}^*}\mathbf{X} = x_1^* x_1 + x_2^* x_2 + \ldots + x_n^* x_n \tag{101}$$
$$= |x_1|^2 + |x_2|^2 + \ldots + |x_n|^2, \tag{102}$$

where bars indicate the moduli of the complex numbers. Similarly the angle between two vectors $\mathbf{X}^{(1)}$ and $\mathbf{X}^{(2)}$ with complex components is now

$$\cos\theta = \frac{\widetilde{\mathbf{X}^{(1)*}}\mathbf{X}^{(2)}}{\{\widetilde{\mathbf{X}^{(1)*}}\mathbf{X}^{(1)}\}^{\frac{1}{2}}\{\widetilde{\mathbf{X}^{(2)*}}\mathbf{X}^{(2)}\}^{\frac{1}{2}}}, \tag{103}$$

which is to be compared with (85). It can easily be verified that θ is an invariant under the unitary transformation $\mathbf{Y} = \mathbf{AX}$ ($\mathbf{A}$ unitary).

Example 4. The matrix

$$\mathbf{A} = \begin{pmatrix} \frac{1}{\sqrt{2}} & \frac{i}{\sqrt{2}} \\ -\frac{i}{\sqrt{2}} & -\frac{1}{\sqrt{2}} \end{pmatrix} \tag{104}$$

is unitary, since

$$\mathbf{A}^{-1} = \begin{pmatrix} \frac{1}{\sqrt{2}} & \frac{i}{\sqrt{2}} \\ -\frac{i}{\sqrt{2}} & -\frac{1}{\sqrt{2}} \end{pmatrix} \text{ and } (\widetilde{\mathbf{A}^*}) = \mathbf{A}^\dagger = \begin{pmatrix} \frac{1}{\sqrt{2}} & \frac{i}{\sqrt{2}} \\ -\frac{i}{\sqrt{2}} & -\frac{1}{\sqrt{2}} \end{pmatrix}. \tag{105}$$

Hence $\mathbf{A}^\dagger = \mathbf{A}^{-1}$.

Now suppose $\mathbf{X}$ is a column vector with components x_1 and ix_2, say, where x_1 and x_2 are real numbers. Then

$$\mathbf{X} = \begin{pmatrix} x_1 \\ ix_2 \end{pmatrix}, \tag{106}$$

and the squared length of $\mathbf{X}$ is

$$(\widetilde{\mathbf{X}^*})\mathbf{X} = \mathbf{X}^\dagger\mathbf{X} = x_1^2 + x_2^2. \tag{107}$$

Now carry out an unitary transformation $\mathbf{Y} = \mathbf{AX}$ using the unitary matrix (104). Then the transformed vector

$$\mathbf{Y} = \begin{pmatrix} y_1 \\ y_2 \end{pmatrix}$$

is given by

$$\begin{pmatrix} y_1 \\ y_2 \end{pmatrix} = \frac{1}{\sqrt{2}} \begin{pmatrix} 1 & i \\ -i & -1 \end{pmatrix} \begin{pmatrix} x_1 \\ ix_2 \end{pmatrix} \tag{108}$$

$$= \frac{1}{\sqrt{2}} \begin{pmatrix} x_1 - x_2 \\ -ix_1 - ix_2 \end{pmatrix}, \tag{109}$$

whence

$$y_1 = \frac{1}{\sqrt{2}}(x_1 - x_2), \qquad y_2 = -\frac{i}{\sqrt{2}}(x_1 + x_2). \tag{110}$$

The squared length of the vector **Y** is

$$\widetilde{\mathbf{Y}^*}\mathbf{Y} = |y_1|^2 + |y_2|^2 = \tfrac{1}{2}(x_1 - x_2)^2 + \tfrac{1}{2}(x_1 + x_2)^2 = x_1^2 + x_2^2, \tag{111}$$

showing that distance is unaltered under an unitary transformation.

PROBLEMS 3

1. Prove that

   (i) adj $\mathbf{AB}$ = adj $\mathbf{B}$ adj $\mathbf{A}$;
   (ii) if $\mathbf{A}$ is symmetric so is adj $\mathbf{A}$;
   (iii) if $\mathbf{A}$ is Hermitian so is adj $\mathbf{A}$;
   (iv) adj (adj $\mathbf{A}$) = $|\mathbf{A}|^{n-2}\mathbf{A}$ if $|\mathbf{A}| \neq 0$.

2. By solving the equations

$$y_1 = x_1 \cos\theta + x_2 \sin\theta$$
$$y_2 = -x_1 \sin\theta + x_2 \cos\theta$$

   for x_1 and x_2, show that the inverse of the matrix

$$\mathbf{A} = \begin{pmatrix} \cos\theta & \sin\theta \\ -\sin\theta & \cos\theta \end{pmatrix} \quad \text{is} \quad \mathbf{A}^{-1} = \begin{pmatrix} \cos\theta & -\sin\theta \\ \sin\theta & \cos\theta \end{pmatrix}.$$

   Verify that $\mathbf{AA}^{-1} = \mathbf{A}^{-1}\mathbf{A} = \mathbf{I}$, and that $\mathbf{A}$ is an orthogonal matrix.

3. Find the inverse of

$$\begin{pmatrix} a+ib & c+id \\ -c+id & a-ib \end{pmatrix}$$

   given that $a^2+b^2+c^2+d^2 = 1$.

4. If
$$\mathbf{A} = \begin{pmatrix} 0 & 0 & 1 \\ 0 & 1 & 0 \\ 1 & 0 & 0 \end{pmatrix},$$
show that $\mathbf{A}^{-1} = \mathbf{A}$.
(Such matrices are called self-reciprocal.)

5. Find the inverse matrix of
$$\mathbf{A} = \begin{pmatrix} 2 & 3 & 1 \\ 3 & 5 & 2 \\ 0 & 0 & 2 \end{pmatrix}$$
and verify that $\mathbf{A}\mathbf{A}^{-1} = \mathbf{A}^{-1}\mathbf{A} = \mathbf{I}$.

6. Verify that
$$\begin{pmatrix} 1 & \alpha & 0 \\ 0 & 1 & 0 \\ 0 & \beta & 1 \end{pmatrix}^{-1} = \begin{pmatrix} 1 & -\alpha & 0 \\ 0 & 1 & 0 \\ 0 & -\beta & 1 \end{pmatrix},$$
and that
$$\begin{pmatrix} 1 & 1 & 0 \\ 0 & 1 & 1 \\ 0 & 0 & 1 \end{pmatrix}^{-1} = \begin{pmatrix} 1 & -1 & 1 \\ 0 & 1 & -1 \\ 0 & 0 & 1 \end{pmatrix}.$$

7. If
$$\alpha + i\beta \equiv \begin{pmatrix} \alpha & \beta \\ -\beta & \alpha \end{pmatrix}$$
verify that
$$(\alpha + i\beta)^{-1} \equiv \begin{pmatrix} \alpha & \beta \\ -\beta & \alpha \end{pmatrix}^{-1}.$$

8. Prove that $(\text{adj } \mathbf{A})^{-1} = (\text{adj } \mathbf{A}^{-1})$.

9. If $\mathbf{A}$ is an $(m \times n)$ matrix with $\tilde{\mathbf{A}}\mathbf{A}$ non-singular, and if $\mathbf{B} = \mathbf{I} - \mathbf{A}(\tilde{\mathbf{A}}\mathbf{A})^{-1}\tilde{\mathbf{A}}$, show that $\mathbf{B} = \mathbf{B}^2$.

10. Obtain by partitioning the inverse of
$$\begin{pmatrix} 1 & 2 & 3 \\ 2 & 4 & 5 \\ 3 & 5 & 6 \end{pmatrix}.$$

11. If
$$\mathbf{X} = \begin{pmatrix} -\frac{1}{2} & -\frac{\sqrt{3}}{2} & 0 \\ -\frac{\sqrt{3}}{2} & \frac{1}{2} & 0 \\ 0 & 0 & 2 \end{pmatrix}$$

and

$$\mathbf{P} = \begin{pmatrix} \frac{1}{2} & \frac{\sqrt{3}}{2} & 0 \\ -\frac{\sqrt{3}}{2} & \frac{1}{2} & 0 \\ 0 & 0 & 1 \end{pmatrix}$$

prove that $\mathbf{P}^{-1}\mathbf{X}\mathbf{P}$ is a diagonal matrix and that $\mathbf{X}$ satisfies the equation

$$\mathbf{X}^3 - 2\mathbf{X}^2 - \mathbf{X} + 2\mathbf{I} = 0,$$

where $\mathbf{I}$ is the unit matrix of order 3.

12. If $\mathbf{A}$ is skew-symmetric (i.e. $\mathbf{A} = -\tilde{\mathbf{A}}$), show that

$$(\mathbf{I}-\mathbf{A})(\mathbf{I}+\mathbf{A})^{-1}$$

is orthogonal (assuming that $\mathbf{I}+\mathbf{A}$ is non-singular).

13. Verify that

$$\begin{pmatrix} \sin\theta & \cos\theta \\ \cos\theta & -\sin\theta \end{pmatrix}$$

is orthogonal.

14. If $\mathbf{A}_n$ is an nth order orthogonal matrix, show that

$$\mathbf{A}_{n+1} = \left(\begin{array}{c|ccccc} 1 & 0 & 0 & \cdots & 0 \\ \hline 0 & & & & \\ 0 & & & \mathbf{A}_n & \\ \vdots & & & & \\ 0 & & & & \end{array}\right)$$

is orthogonal.

15. If $\mathbf{AB} = \mathbf{BA}$, show that $\mathbf{RA}\tilde{\mathbf{R}}$ and $\mathbf{RB}\tilde{\mathbf{R}}$ commute if $\mathbf{R}$ is orthogonal.

16. Show that

$$\mathbf{S} = \begin{pmatrix} \frac{1}{\sqrt{2}} & -\frac{1}{\sqrt{2}} \\ \frac{1}{\sqrt{2}} & \frac{1}{\sqrt{2}} \end{pmatrix}$$

is orthogonal. Show also that if

$$\mathbf{P} = \begin{pmatrix} 1 & 3 \\ 3 & 1 \end{pmatrix}$$

then $\mathbf{SP\tilde{S}}$ is a diagonal matrix.

17. Prove that if $\mathbf{A}$ is skew-Hermitian then

$$(\mathbf{I}-\mathbf{A})(\mathbf{I}+\mathbf{A})^{-1}$$

is unitary (assuming that $\mathbf{I}+\mathbf{A}$ is non-singular).

18. Show that

$$\left|\begin{pmatrix} \mathbf{A} & \mathbf{B} \\ 0 & \mathbf{C} \end{pmatrix}\right| = |\mathbf{A}|\,|\mathbf{C}|,$$

where $\mathbf{A}$ and $\mathbf{C}$ are square matrices, and $\mathbf{A}^{-1}$ exists. By considering the matrix product

$$\begin{pmatrix} \mathbf{I}+\mathbf{A\tilde{B}} & 0 \\ \mathbf{\tilde{B}} & 1 \end{pmatrix}\begin{pmatrix} \mathbf{I} & \mathbf{A} \\ 0 & 1 \end{pmatrix}$$

show that

$$|\mathbf{I}+\mathbf{A\tilde{B}}| = \mathbf{\tilde{B}A}+1.$$

CHAPTER 4

Systems of Linear Algebraic Equations

4.1 Introduction

One of the important uses of matrices occurs in the solution of systems of linear algebraic equations. However, this is almost a subject in its own right since the problem really reduces to the numerical computation of inverse matrices. Various numerical procedures are now available for the inversion of large matrices to any desired degree of accuracy, and the reader who is interested in these techniques should consult one or other of the treatises listed at the end of this book (in particular, a useful account is given in the text by Fox, *An Introduction to Numerical Linear Algebra*). For this reason only a somewhat formal and elementary account of the solution of linear equations is given here.

4.2 Non-homogeneous equations

Consider the set of n linear algebraic equations in n unknowns $x_1, x_2, \ldots, x_n$

$$\left.\begin{array}{c} a_{11}x_1+a_{12}x_2+\ldots+a_{1n}x_n = h_1, \\ a_{21}x_1+a_{22}x_2+\ldots+a_{2n}x_n = h_2, \\ \cdot \quad \cdot \quad \cdot \quad \cdot \\ \cdot \quad \cdot \quad \cdot \quad \cdot \\ \cdot \quad \cdot \quad \cdot \quad \cdot \\ a_{n1}x_1+a_{n2}x_2+\ldots+a_{nn}x_n = h_n, \end{array}\right\} \tag{1}$$

where the coefficients a_{ik} and h_i $(i, k = 1, 2, \ldots, n)$ are known constants. Writing (1) in matrix form, we have

$$\mathbf{AX} = \mathbf{H}, \tag{2}$$

where $$\mathbf{A} = \begin{pmatrix} a_{11} & a_{12} & \ldots & a_{1n} \\ a_{21} & a_{22} & \ldots & a_{2n} \\ \cdot & & & \cdot \\ \cdot & & & \cdot \\ \cdot & & & \cdot \\ a_{n1} & \cdot & \cdot & a_{nn} \end{pmatrix}, \quad \mathbf{X} = \begin{pmatrix} x_1 \\ x_2 \\ \cdot \\ \cdot \\ \cdot \\ x_n \end{pmatrix} \tag{3}$$

and

$$\mathbf{H} = \begin{pmatrix} h_1 \\ h_2 \\ \cdot \\ \cdot \\ \cdot \\ h_n \end{pmatrix}. \tag{4}$$

Provided **H** is a non-zero column vector (i.e. not all elements equal to zero), equations (1) are called non-homogeneous. Homogeneous equations ($\mathbf{H} = \mathbf{0}$) will be discussed later in this chapter.

Now returning to (2) and assuming that **A** has an inverse $\mathbf{A}^{-1}$ we have, by pre-multiplying by $\mathbf{A}^{-1}$ throughout,

$$\mathbf{A}^{-1}\mathbf{A}\mathbf{X} = \mathbf{A}^{-1}\mathbf{H}, \tag{5}$$

which, since $\mathbf{A}^{-1}\mathbf{A} = \mathbf{I}$, gives

$$\mathbf{X} = \mathbf{A}^{-1}\mathbf{H}. \tag{6}$$

This matrix equation gives the solution of (2).

Example 1. To solve the equations

$$\left.\begin{aligned} x+y+z &= 6, \\ x+2y+3z &= 14, \\ x+4y+9z &= 36. \end{aligned}\right\} \tag{7}$$

Now

$$\mathbf{A} = \begin{pmatrix} 1 & 1 & 1 \\ 1 & 2 & 3 \\ 1 & 4 & 9 \end{pmatrix}, \quad \mathbf{X} = \begin{pmatrix} x \\ y \\ z \end{pmatrix} \quad \text{and} \quad \mathbf{H} = \begin{pmatrix} 6 \\ 14 \\ 36 \end{pmatrix}. \tag{8}$$

Hence

$$\mathbf{A}^{-1} = \frac{\text{adj } \mathbf{A}}{|\mathbf{A}|} = \tfrac{1}{2}\begin{pmatrix} 6 & -5 & 1 \\ -6 & 8 & -2 \\ 2 & -3 & 1 \end{pmatrix}, \tag{9}$$

and consequently, since $\mathbf{X} = \mathbf{A}^{-1}\mathbf{H}$,

$$\begin{pmatrix} x \\ y \\ z \end{pmatrix} = \begin{pmatrix} 3 & -\frac{5}{2} & \frac{1}{2} \\ -3 & 4 & -1 \\ 1 & -\frac{3}{2} & \frac{1}{2} \end{pmatrix} \begin{pmatrix} 6 \\ 14 \\ 36 \end{pmatrix} = \begin{pmatrix} 1 \\ 2 \\ 3 \end{pmatrix} \tag{10}$$

The solutions are therefore $x = 1$, $y = 2$ and $z = 3$.

The general matrix method just outlined may be used to derive Cramer's rule for the solution of linear equations by determinants. For from (6) we have

$$\mathbf{X} = \mathbf{A}^{-1}\mathbf{H} = \frac{1}{|\mathbf{A}|}(\text{adj } \mathbf{A})\mathbf{H}, \tag{11}$$

which, using the definition of adj **A** given in Chapter 3, 3.2, gives

$$\left.\begin{aligned}
x_1 &= \frac{1}{|\mathbf{A}|}(h_1 A_{11} + h_2 A_{21} + \ldots + h_n A_{n1}), \\
x_2 &= \frac{1}{|\mathbf{A}|}(h_1 A_{12} + h_2 A_{22} + \ldots + h_n A_{n2}), \\
&\;\;\vdots \\
x_r &= \frac{1}{|\mathbf{A}|}(h_1 A_{1r} + h_2 A_{2r} + \ldots + h_n A_{nr}), \\
&\;\;\vdots \\
x_n &= \frac{1}{|\mathbf{A}|}(h_1 A_{1n} + h_2 A_{2n} + \ldots + h_n A_{nn}).
\end{aligned}\right\} \tag{12}$$

Each of these expressions is the expansion of a determinant divided by $|\mathbf{A}|$. It is easily seen that in fact

$$x_1 = \frac{\begin{vmatrix} h_1 & a_{12} & \ldots & a_{1n} \\ h_2 & a_{22} & \ldots & \cdot \\ \cdot & \cdot & & \cdot \\ \cdot & \cdot & & \cdot \\ \cdot & \cdot & & \cdot \\ h_n & a_{n2} & \ldots & a_{nn} \end{vmatrix}}{\begin{vmatrix} a_{11} & a_{12} & \ldots & a_{1n} \\ a_{21} & a_{22} & \ldots & a_{2n} \\ \cdot & & & \cdot \\ \cdot & & & \\ \cdot & & & \\ a_{n1} & \ldots & \ldots & a_{nn} \end{vmatrix}}, \quad x_2 = \frac{\begin{vmatrix} a_{11} & h_1 & \ldots & a_{1n} \\ a_{21} & h_2 & \ldots & a_{2n} \\ \cdot & \cdot & & \cdot \\ \cdot & \cdot & & \cdot \\ \cdot & \cdot & & \cdot \\ a_{n1} & h_n & \ldots & a_{nn} \end{vmatrix}}{\begin{vmatrix} a_{11} & a_{12} & \ldots & a_{1n} \\ a_{21} & a_{22} & \ldots & a_{2n} \\ & & & \cdot \\ & & & \cdot \\ & & & \cdot \\ a_{n1} & \ldots & \ldots & a_{nn} \end{vmatrix}}, \tag{13}$$

and so on. In general we have

$$x_r = \frac{\begin{vmatrix} a_{11} & a_{12} & \dots & h_1 & \dots & a_{1n} \\ a_{21} & a_{22} & \dots & h_2 & \dots & a_{2n} \\ \cdot & \cdot & & \cdot & & \cdot \\ \cdot & \cdot & & \cdot & & \cdot \\ \cdot & \cdot & & \cdot & & \cdot \\ a_{n1} & a_{n2} & \dots & h_n & \dots & a_{nn} \end{vmatrix}}{\begin{vmatrix} a_{11} & a_{12} & \dots & a_{1r} & \dots & a_{1n} \\ a_{21} & a_{22} & \dots & a_{2r} & \dots & a_{2n} \\ \cdot & \cdot & & \cdot & & \cdot \\ \cdot & \cdot & & \cdot & & \cdot \\ \cdot & \cdot & & \cdot & & \cdot \\ a_{n1} & a_{n2} & \dots & a_{nr} & \dots & a_{nn} \end{vmatrix}} \quad (r = 1, 2, \dots, n), \qquad (14)$$

where the determinant in the numerator is obtained from the determinant in the denominator (i.e. $|\mathbf{A}|$) by replacing the rth column by the elements $h_1, h_2, \dots, h_n$. This is Cramer's rule. As an example we take the equations of Example 1 again. Using (14) we have

$$x_1 = \frac{\begin{vmatrix} 6 & 1 & 1 \\ 14 & 2 & 3 \\ 36 & 4 & 9 \end{vmatrix}}{\begin{vmatrix} 1 & 1 & 1 \\ 1 & 2 & 3 \\ 1 & 4 & 9 \end{vmatrix}} = 1, \qquad x_2 = \frac{\begin{vmatrix} 1 & 6 & 1 \\ 1 & 14 & 3 \\ 1 & 36 & 9 \end{vmatrix}}{\begin{vmatrix} 1 & 1 & 1 \\ 1 & 2 & 3 \\ 1 & 4 & 9 \end{vmatrix}} = 2, \qquad (15)$$

and

$$x_3 = \frac{\begin{vmatrix} 1 & 1 & 6 \\ 1 & 2 & 14 \\ 1 & 3 & 36 \end{vmatrix}}{\begin{vmatrix} 1 & 1 & 1 \\ 1 & 2 & 3 \\ 1 & 4 & 9 \end{vmatrix}} = 3, \qquad (16)$$

where $x_1 \equiv x$, $x_2 \equiv y$ and $x_3 \equiv z$.

Now from the method outlined by equations (2)–(6) it is clear that a solution of n linear algebraic equations in n unknowns exists provided $\mathbf{A}^{-1}$ exists – that is, provided $|\mathbf{A}| \neq 0$. This solution is in fact unique. For suppose the solution is $\mathbf{X} = \mathbf{X}_1$ so that $\mathbf{AX}_1 = \mathbf{H}$. Let $\mathbf{X}_2$ be another solution. Then $\mathbf{AX}_2 = \mathbf{H}$. Consequently

$\mathbf{AX}_1 = \mathbf{AX}_2$. But, since $|\mathbf{A}| \neq 0$, $\mathbf{X}_1 = \mathbf{X}_2$, and hence the solution is unique.

We must now consider the two possibilities which can arise when $|\mathbf{A}| = 0$.

(a) If any of the determinants in the numerators of (14) are non-zero, then, since the determinant in the denominator (i.e. $|\mathbf{A}|$) is zero, no finite solution of the set of equations exists. The equations are then said to be inconsistent or incompatible.

For example, the equations

$$\left.\begin{aligned} 3x+2y &= 2, \\ 3x+2y &= 6, \end{aligned}\right\} \tag{17}$$

are of this type since, by Cramer's rule,

$$x = \frac{\begin{vmatrix} 2 & 2 \\ 6 & 2 \end{vmatrix}}{\begin{vmatrix} 3 & 2 \\ 3 & 2 \end{vmatrix}} = \frac{-8}{0}, \quad \text{and} \quad y = \frac{\begin{vmatrix} 3 & 2 \\ 3 & 6 \end{vmatrix}}{\begin{vmatrix} 3 & 2 \\ 3 & 2 \end{vmatrix}} = \frac{12}{0}, \tag{18}$$

which are not defined quantities. No finite solution exists therefore. This result may be interpreted geometrically by noting that (18) represents two non-intersecting straight lines.

Similarly the set of equations

$$\left.\begin{aligned} -2x+y+z &= 1, \\ x-2y+z &= 2, \\ x+y-2z &= 3, \end{aligned}\right\} \tag{19}$$

(for which $|\mathbf{A}| = 0$) has no finite solution. The equations are inconsistent since the negative of the sum of the last two equations gives $-2x+y+z = -5$, which is inconsistent with the first equation of the set.

(b) If, in addition to $|\mathbf{A}| = 0$, the determinants in the numerators of (14) are all zero, then in general an infinity of solutions exist. For example, the equations

$$\left.\begin{aligned} 3x+2y &= 2, \\ 6x+4y &= 4, \end{aligned}\right\} \tag{20}$$

are of this type. The second equation is just the first equation in disguise. Consequently there is only one equation for two unknowns with the result that an infinity of (x, y) values satisfy the equation.

The equations are linearly dependent in that one is a multiple of the other. (Geometrically (20) represents two coincident lines with an infinity of common points.) Similarly the equations

$$\left.\begin{aligned} -2x+y+z &= 1,\\ x-2y+z &= 2,\\ x+y-2z &= -3, \end{aligned}\right\} \tag{21}$$

have $|\mathbf{A}| = 0$, and all numerator determinants equal to zero. Again the equations are linearly dependent in that the first is just the negative of the sum of the second and third and is therefore redundant. Consequently (21) is in reality only a pair of equations for three unknowns x, y and z – namely

$$\left.\begin{aligned} x-2y+z &= 2,\\ x+y-2z &= -3, \end{aligned}\right\} \tag{22}$$

which are satisfied by the infinity of solutions of the form

$$x = \lambda-\tfrac{4}{3}, \qquad y = \lambda-\tfrac{5}{3}, \qquad z = \lambda, \tag{23}$$

where λ is an arbitrary parameter.

4.3 Homogeneous equations

We now consider a system of n homogeneous equations

$$\left.\begin{aligned} a_{11}x_1+a_{12}x_2+\ldots+a_{1n}x_n &= 0,\\ a_{21}x_1+a_{22}x_2+\ldots+a_{2n}x_n &= 0,\\ \vdots \qquad \vdots \qquad & \vdots\\ a_{n1}x_1+a_{n2}x_2+\ldots+a_{nn}x_n &= 0, \end{aligned}\right\} \tag{24}$$

which is obtained by putting the elements $h_1, h_2, \ldots, h_n$ in (1) equal to zero. In matrix form, therefore, (24) becomes

$$\mathbf{AX} = \mathbf{0}, \tag{25}$$

where $\mathbf{0}$ is the zero column matrix (or vector) of order n. If $|\mathbf{A}| \neq 0$ then $\mathbf{A}^{-1}$ exists and consequently by (6)

$$\mathbf{X} = \mathbf{A}^{-1}\mathbf{0} = \mathbf{0} \tag{26}$$

is the only solution – that is, $x_1 = 0$, $x_2 = 0$, $\ldots$, $x_n = 0$. This identically zero solution is usually called the trivial solution and is of little interest. However, if $|\mathbf{A}| = 0$ an infinity of non-trivial solutions exists as in the non-homogeneous case.

For example, the set

$$\left.\begin{aligned} x+5y+3z &= 0, \\ 5x+y-kz &= 0, \\ x+2y+kz &= 0, \end{aligned}\right\} \tag{27}$$

where k is an arbitrary parameter, has the trivial solution $x = y = z = 0$ for all values of k. Non-trivial solutions will exist, however, when

$$|\mathbf{A}| = \begin{vmatrix} 1 & 5 & 3 \\ 5 & 1 & -k \\ 1 & 2 & k \end{vmatrix} = 27(1-k)=0, \tag{28}$$

which gives $k = 1$. In this case the equations are linearly dependent since

$$x+5y+3z = -\tfrac{1}{3}(5x+y-z)+\tfrac{8}{3}(x+2y+z). \tag{29}$$

Consequently there are only two equations for three unknowns. Solving any two of (27) we find the infinity of solutions

$$x = -\lambda, \qquad y = 2\lambda, \qquad z = -3\lambda, \tag{30}$$

where λ is an arbitrary parameter.

4.4 Ill-conditioned equations

In many instances where the set of equations of type (1) arise from the mathematical description of an experimental set-up, the coefficients a_{ik}, h_i $(i, k = 1, 2, \ldots, n)$ may be known only approximately as experimentally determined values subject to certain errors. If, in addition, the value of $|\mathbf{A}|$ is small compared with the magnitude of the coefficients a_{ik}, h_i then the solution of the set of equations may be very sensitive to small changes in the values of the coefficients. Such equations are called ‘ ill-conditioned ’. Consider, for example, the pair of equations

$$\left.\begin{aligned} 3x+1{\cdot}52y &= 1, \\ 2x+1{\cdot}02y &= 1, \end{aligned}\right\} \tag{31}$$

for which

$$|\mathbf{A}| = \begin{vmatrix} 3 & 1{\cdot}52 \\ 2 & 1{\cdot}02 \end{vmatrix} = 0{\cdot}02. \tag{32}$$

By Cramer's rule, the solution of (31) is

$$x = -25, \qquad y = 50. \tag{33}$$

Now allow a small change in the a_{22} coefficient so that the equations read, for example,

$$\left.\begin{aligned} 3x+1{\cdot}52y &= 1, \\ 2x+1{\cdot}03y &= 1. \end{aligned}\right\} \tag{34}$$

Then $|\mathbf{A}| = 0{\cdot}05$ and, by Cramer's rule again,

$$x = -9{\cdot}8, \qquad y = 20. \tag{35}$$

Clearly if the experimental design leads to a set of ill-conditioned equations no reliance can be placed on the solutions of such equations for, as the example shows, a change of about 1% in one of the coefficients can lead to a change of some 200%–300% in the solutions. In general, there is no mathematical way of overcoming this difficulty. The best that can be done is to re-interpret the experimental set-up (using different variables, for example) in an attempt to obtain a set of equations which is not ill-conditioned.

PROBLEMS 4

1. Solve by matrix methods the equations

$$\begin{aligned} 4x-3y+z &= 11, \\ 2x+y-4z &= -1, \\ x+2y-2z &= 1. \end{aligned}$$

2. Show that the three equations

$$\begin{aligned} -2x+y+z &= a, \\ x-2y+z &= b, \\ x+y-2z &= c, \end{aligned}$$

have no solutions unless $a+b+c = 0$, in which case they have infinitely many. Find these solutions when $a = 1, b = 1, c = -2$.

3. Show that there are two values of k for which the equations

$$\begin{aligned} kx+3y+2z &= 1, \\ x+(k-1)y &= 4, \\ 10y+3z &= -2, \\ 2x-ky-z &= 5, \end{aligned}$$

are consistent. Find their common solution for that value of k which is an integer.

4. Solve, where possible, the following sets of equations:

(a)
$$\begin{aligned} 8x-4y+z &= 8, \\ 4x+2y-2z &= 0, \\ 2x+7y-4z &= 0. \end{aligned}$$

(b)
$$\begin{aligned} x+y-z+w &= 0, \\ 3x-y+2z+3w &= 7, \\ x+2y-2z-w &= -1, \\ 3z+w &= 9. \end{aligned}$$

(c)
$$\begin{aligned} x-2y+3z &= 0, \\ 2x+5y+6z &= 0. \end{aligned}$$

(d)
$$\begin{aligned} x+y+z &= 1, \\ x-y+2z &= 5, \\ 3x+y+z &= 2, \\ 2x-2y+3z &= 1. \end{aligned}$$

(e)
$$\begin{aligned} x+y+z+w &= 1, \\ 2x-y+z-2w &= 2, \\ 3x+2y-z-w &= 3. \end{aligned}$$

(f)
$$\begin{aligned} x+5y+3z &= 1, \\ 5x+y-z &= 2, \\ x+2y+z &= 3. \end{aligned}$$

(g)
$$\begin{aligned} x+y-z &= 0, \\ 2x+3y-3z &= 1, \\ -x+4y-z &= 3, \\ 4x-5y+z &= 2. \end{aligned}$$

5. Suppose $\mathbf{B}$ is good approximation to the inverse $\mathbf{A}^{-1}$ of $\mathbf{A}$. Let $\mathbf{AB} = \mathbf{I}-\delta\mathbf{E}$, so that $\delta\mathbf{E} = \mathbf{0}$ when $\mathbf{B} = \mathbf{A}^{-1}$. Deduce that

$$\mathbf{A}^{-1} = \mathbf{B}+\mathbf{A}^{-1}\,\delta\mathbf{E},$$

and that

$$\mathbf{A}^{-1} = \mathbf{B}+(\mathbf{B}+\mathbf{A}^{-1}\,\delta\mathbf{E})\,\delta\mathbf{E} = \mathbf{B}(\mathbf{I}+\delta\mathbf{E}+\delta\mathbf{E}^2+\delta\mathbf{E}^3+\ldots).$$

If the elements of $\delta\mathbf{E}$ are sufficiently small the series will converge and $\mathbf{A}^{-1}$ may be calculated to any desired degree of accuracy. Use this result (up to the $\delta\mathbf{E}^3$ term) to obtain an improved approximation to $\mathbf{A}^{-1}$, where

$$\mathbf{A} = \begin{pmatrix} 4 & 2 \\ 1 & 3 \end{pmatrix} \quad \text{and} \quad \mathbf{B} = \begin{pmatrix} \frac{1}{2} & -\frac{1}{4} \\ -\frac{1}{4} & \frac{1}{2} \end{pmatrix}.$$

6. Obtain the solutions of

$$x+1{\cdot}52y = 1,$$
$$2x+(3{\cdot}05+\delta)y = 1,$$

for $\delta = -0{\cdot}02,\ -0{\cdot}01,\ 0,\ 0{\cdot}01$ and $0{\cdot}02$.

7. Consider the following m linear equations in n unknowns $x_1, x_2, \ldots, x_n$ where $m > n$:

$$\begin{aligned} a_{11}x_1+a_{12}x_2+\ldots+a_{1n}x_n &= h_1, \\ a_{21}x_1+a_{22}x_2+\ldots+a_{2n}x_n &= h_2, \\ \vdots \qquad \qquad \qquad & \\ a_{m1}x_1+a_{m2}x_2+\ldots+a_{mn}x_n &= h_m. \end{aligned}$$

In general these equations will be inconsistent. The ‘ best values ’ of $x_1, x_2, \ldots, x_n$ are then defined as those which make the expression

$$E = \sum_{r=1}^{m} \{a_{r1}x_1+a_{r2}x_2+\ldots+a_{rn}x_n-h_r\}^2$$

a minimum. The necessary conditions for this to be so are

$$\frac{\partial E}{\partial x_1} = \frac{\partial E}{\partial x_2} = \ldots = \frac{\partial E}{\partial x_n} = 0.$$

Show that these conditions lead to the set of n equations (the ‘ normal ’ equations)

$$\sum_{r=1}^{m} a_{r1}(a_{r1}x_1+a_{r2}x_2+\ldots+a_{rn}x_n-h_r)=0$$

$$\sum_{r=1}^{m} a_{r2}(a_{r1}x_1+a_{r2}x_2+\ldots+a_{rn}x_n-h_r)=0$$

$$\vdots \qquad \vdots \qquad \vdots \qquad \vdots$$

$$\sum_{r=1}^{m} a_{rn}(a_{r1}x_1+a_{r2}x_2+\ldots+a_{rn}x_n-h_r)=0$$

for the best values of $x_1, x_2, \ldots, x_n$.
(This is the method of least squares.)

Find the best values of x and y for the inconsistent equations

$$2x+3y=8,$$
$$3x-y=1,$$
$$x+y=4.$$

CHAPTER 5

Eigenvalues and Eigenvectors

5.1 Introduction

In Chapter 1, 1.6 a typical oscillation problem was formulated in matrix language and it was shown that the equations of motion could be written as (see equation (40), Chapter 1).

$$\ddot{\mathbf{Y}} = \mathbf{AY}, \tag{1}$$

where $\mathbf{Y}$ is a (2×1) column vector and $\mathbf{A}$ a second order square matrix, the dots denoting differentiation with respect to time t. In an attempt to solve (1) we write

$$\mathbf{Y} = \mathbf{X}\, e^{\omega t}, \tag{2}$$

where $\mathbf{X}$ is a column vector independent of t, and where ω is a constant. Equation (1) now becomes

$$\mathbf{AX} = \omega^2 \mathbf{X}, \tag{3}$$

which may be written as a set of homogeneous equations

$$(\mathbf{A} - \lambda \mathbf{I})\,\mathbf{X} = \mathbf{0}, \tag{4}$$

where $\lambda = \omega^2$ and $\mathbf{I}$ is the unit matrix of the same order as $\mathbf{A}$.

Besides having the trivial solution $\mathbf{X} = \mathbf{0}$ (which is of no interest), equation (4) will have non-trivial solutions ($\mathbf{X} \neq \mathbf{0}$) only if

$$|\mathbf{A} - \lambda \mathbf{I}| = 0 \tag{5}$$

(see Chapter 4, 4.3). This equation is called the characteristic equation of $\mathbf{A}$ and determines those values of λ for which non-trivial solutions of (4) will exist. In the particular oscillation problem considered here (since $\lambda = \omega^2$) the λ values determine the frequencies of oscillation ω. Equations of the type (4) arise frequently in the solution of many other types of physical problems and in the sections which follow therefore we discuss the general nature of these equations and their solutions.

5.2 Eigenvalues and eigenvectors

Suppose **A** is a square matrix of order n with elements a_{ik}, and **X** is a column vector of order n with elements x_i. Then the set of homogeneous equations

$$(\mathbf{A}-\lambda\mathbf{I})\,\mathbf{X}=\mathbf{0} \tag{6}$$

has non-trivial solutions only if

$$|\mathbf{A}-\lambda\mathbf{I}|=0. \tag{7}$$

That is, only if

$$\begin{vmatrix} a_{11}-\lambda & a_{12} & \dots & a_{1n} \\ a_{21} & a_{22}-\lambda & \dots & a_{2n} \\ \cdot & \cdot & & \cdot \\ \cdot & \cdot & & \cdot \\ \cdot & \cdot & & \cdot \\ a_{n1} & a_{n2} & \dots & a_{nn}-\lambda \end{vmatrix}=0. \tag{8}$$

As already indicated in 5.1, equation (7) (or (8)) is called the characteristic (or secular) equation of the matrix **A**. The expansion of $|\mathbf{A}-\lambda\mathbf{I}|$ gives rise to an nth degree polynomial in λ, say $f(\lambda)$, called the characteristic polynomial, and the roots $\lambda_1, \lambda_2, \ldots, \lambda_n$ of the characteristic equation $f(\lambda) = 0$ are called the eigenvalues (or characteristic roots, latent roots or proper values) of the matrix **A**. To each root $\lambda_i (i = 1, 2, \ldots, n)$ there is a non-trivial solution $\mathbf{X}_i$ called the eigenvector (or characteristic vector or latent vector). For any other value of $\lambda \neq \lambda_i$ the only solution of (6) is the trivial one $\mathbf{X} = \mathbf{0}$.

Finally it is important to note that, since the set of equations (6) is homogeneous, if $\mathbf{X}_i$ is an eigenvector belonging to an eigenvalue λ_i then so also is $k\mathbf{X}_i$, where k is an arbitrary non-zero constant. The length of the eigenvector is therefore undetermined by the equations. If the elements of the eigenvectors are all real then the length of $k\mathbf{X}_i$ is given by

$$\{(\widetilde{k\mathbf{X}_i})k\mathbf{X}_i\}^{\frac{1}{2}} = k\{\widetilde{\mathbf{X}}_i\mathbf{X}_i\}^{\frac{1}{2}}, \tag{9}$$

and it is usual to choose k such that the eigenvector has unit length. (If some of the elements of the eigenvector are complex then the length of $k\mathbf{X}_i$ is defined as $k\{(\widetilde{\mathbf{X}_i^*})\mathbf{X}_i\}^{\frac{1}{2}}$ – see Chapter 3, 3.6.) In most of what follows eigenvectors will be normalised to unit length.

Example 1. To find the eigenvalues and eigenvectors of the matrix

$$\mathbf{A} = \begin{pmatrix} 4 & 1 \\ 2 & 3 \end{pmatrix}. \tag{10}$$

Now the characteristic equation is

$$|\mathbf{A}-\lambda\mathbf{I}| = \begin{vmatrix} 4-\lambda & 1 \\ 2 & 3-\lambda \end{vmatrix} = (\lambda-2)(\lambda-5) = 0. \tag{11}$$

Hence the two eigenvalues are $\lambda_1 = 2$, $\lambda_2 = 5$. To find the eigenvectors belonging to these two eigenvalues we take the basic set of homogeneous equations (6) with $\mathbf{A}$ given by (10) and solve for $\mathbf{X}$ for each λ value. In general, the eigenvector corresponding to the ith eigenvalue λ_i will be denoted by $\mathbf{X}_i$ and the elements (or components) of $\mathbf{X}_i$ denoted by $x_1{}^{(i)}, x_2{}^{(i)}, \ldots, x_n{}^{(i)}$.

Case $\lambda_1 = 2$

Equations (6) become

$$\left[\begin{pmatrix} 4 & 1 \\ 2 & 3 \end{pmatrix} - 2\begin{pmatrix} 1 & 0 \\ 0 & 1 \end{pmatrix}\right]\begin{pmatrix} x_1^{(1)} \\ x_2^{(1)} \end{pmatrix} = \begin{pmatrix} 0 \\ 0 \end{pmatrix}. \tag{12}$$

Hence

$$2x_1^{(1)} + x_2^{(1)} = 0 \quad \text{(twice)}, \tag{13}$$

giving

$$x_1^{(1)} = -\tfrac{1}{2}x_2^{(1)}. \tag{14}$$

If we normalise the length of the eigenvector to unity we require $x_1^{(1)2} + x_2^{(1)2} = 1$, so that using (14)

$$x_1^{(1)} = -\frac{1}{\sqrt{5}}, \qquad x_2^{(1)} = \frac{2}{\sqrt{5}}. \tag{15}$$

Consequently the normalised eigenvector corresponding to $\lambda_1 = 2$ is

$$\mathbf{X}_1 = \begin{pmatrix} -\dfrac{1}{\sqrt{5}} \\ \dfrac{2}{\sqrt{5}} \end{pmatrix}. \tag{16}$$

Case $\lambda_2 = 5$

Equations (6) now become

$$\left[\begin{pmatrix} 4 & 1 \\ 2 & 3 \end{pmatrix} - 5\begin{pmatrix} 1 & 0 \\ 0 & 1 \end{pmatrix}\right]\begin{pmatrix} x_1^{(2)} \\ x_2^{(2)} \end{pmatrix} = \begin{pmatrix} 0 \\ 0 \end{pmatrix}, \tag{17}$$

which are

$$\left.\begin{aligned} -x_1^{(2)}+x_2^{(2)} &= 0, \\ 2x_1^{(2)}-2x_2^{(2)} &= 0, \end{aligned}\right\} \tag{18}$$

giving

$$x_1^{(2)} = x_2^{(2)}. \tag{19}$$

Normalising to unit length as before so that $x_1^{(2)^2}+x_2^{(2)^2} = 1$ we find

$$x_1^{(2)} = \frac{1}{\sqrt{2}}, \qquad x_2^{(2)} = \frac{1}{\sqrt{2}}. \tag{20}$$

Hence the normalised eigenvector corresponding to the eigenvalue $\lambda_2 = 2$ is

$$\mathbf{X}_2 = \begin{pmatrix} \frac{1}{\sqrt{2}} \\ \frac{1}{\sqrt{2}} \end{pmatrix}. \tag{21}$$

Example 2. To find the eigenvalues and eigenvectors of the matrix

$$\mathbf{A} = \begin{pmatrix} 1 & -1 & -1 \\ 1 & -1 & 0 \\ 1 & 0 & -1 \end{pmatrix}. \tag{22}$$

The characteristic equation is

$$\begin{vmatrix} 1-\lambda & -1 & -1 \\ 1 & -1-\lambda & 0 \\ 1 & 0 & -1-\lambda \end{vmatrix} = 0, \tag{23}$$

which gives three roots $\lambda_1 = -1$, $\lambda_2 = i$, $\lambda_3 = -i$.

Case $\lambda_1 = -1$

Equations (6) are

$$\left[\begin{pmatrix} 1 & -1 & -1 \\ 1 & -1 & 0 \\ 1 & 0 & -1 \end{pmatrix} + 1\begin{pmatrix} 1 & 0 & 0 \\ 0 & 1 & 0 \\ 0 & 0 & 1 \end{pmatrix}\right] \begin{pmatrix} x_1^{(1)} \\ x_2^{(1)} \\ x_3^{(1)} \end{pmatrix} = \begin{pmatrix} 0 \\ 0 \\ 0 \end{pmatrix}, \tag{24}$$

which give

$$\left.\begin{aligned} 2x_1^{(1)}-x_2^{(1)}-x_3^{(1)} &= 0, \\ x_1^{(1)} &= 0 \text{ (twice).} \end{aligned}\right\} \tag{25}$$

Hence

$$x_1^{(1)} = 0, \qquad x_2^{(1)} = -x_3^{(1)}. \tag{26}$$

Normalising to unit length so that $x_1^{(1)^2}+x_2^{(1)^2}+x_3^{(1)^2} = 1$ we have

$$x_1^{(1)} = 0, \qquad x_2^{(1)} = \frac{1}{\sqrt{2}}, \qquad x_3^{(1)} = -\frac{1}{\sqrt{2}}. \tag{27}$$

Consequently

$$\mathbf{X}_1 = \begin{pmatrix} 0 \\ \dfrac{1}{\sqrt{2}} \\ -\dfrac{1}{\sqrt{2}} \end{pmatrix}. \tag{28}$$

Case $\lambda_2 = +i$

Here equations (6) take the form

$$\left[\begin{pmatrix} 1 & -1 & -1 \\ 1 & -1 & 0 \\ 1 & 0 & -1 \end{pmatrix} - i \begin{pmatrix} 1 & 0 & 0 \\ 0 & 1 & 0 \\ 0 & 0 & 1 \end{pmatrix}\right] \begin{pmatrix} x_1^{(2)} \\ x_2^{(2)} \\ x_3^{(2)} \end{pmatrix} = \begin{pmatrix} 0 \\ 0 \\ 0 \end{pmatrix}, \tag{29}$$

giving

$$\left.\begin{aligned} (1-i)x_1^{(2)} - x_2^{(2)} - x_3^{(2)} &= 0, \\ x_1^{(2)} - (1+i)x_2^{(2)} &= 0, \\ x_1^{(2)} - (1+i)x_3^{(2)} &= 0, \end{aligned}\right\} \tag{30}$$

the solutions of which are

$$x_1^{(2)} = (1+i)x_2^{(2)}, \qquad x_2^{(2)} = x_3^{(2)}. \tag{31}$$

Now, since some of the elements of the eigenvector are complex, when normalising to unit length we must use the generalised definition of distance (see Chapter 3, 3.6) and require

$$x_1^{*(2)}x_1^{(2)} + x_2^{*(2)}x_2^{(2)} + x_3^{*(2)}x_3^{(2)} = 1. \tag{32}$$

Consequently

$$x_1^{(2)} = \frac{1+i}{2}, \qquad x_2^{(2)} = \tfrac{1}{2}, \qquad x_3^{(2)} = \tfrac{1}{2}. \tag{33}$$

The normalised eigenvector is therefore

$$\mathbf{X}_2 = \begin{pmatrix} \dfrac{1+i}{2} \\ \tfrac{1}{2} \\ \tfrac{1}{2} \end{pmatrix}. \tag{34}$$

Case $\lambda_3 = -i$

Proceeding in a similar fashion we find in this case

$$x_1^{(3)} = \frac{1-i}{2}, \qquad x_2^{(3)} = \tfrac{1}{2}, \qquad x_3^{(3)} = \tfrac{1}{2}, \tag{35}$$

giving

$$\mathbf{X}_3 = \begin{pmatrix} \frac{1-i}{2} \\ \frac{1}{2} \\ \frac{1}{2} \end{pmatrix}. \tag{36}$$

5.3 Some properties of eigenvalues

We now return to the characteristic equation (8). The left-hand side of this equation may be written

$$|\mathbf{A}-\lambda\mathbf{I}| = f(\lambda) = (-1)^n \{\lambda^n - \alpha_1\lambda^{n-1} + \alpha_2\lambda^{n-2} \ldots + (-1)^n\alpha_n\}, \tag{37}$$

where $\alpha_1, \alpha_2, \ldots, \alpha_n$ are defined in terms of the elements a_{ik}. Suppose now that $\lambda_1, \lambda_2, \ldots, \lambda_n$ are the n roots (the eigenvalues) of the characteristic equation $f(\lambda) = 0$. Then

$$f(\lambda) = (\lambda_1-\lambda)(\lambda_2-\lambda)(\lambda_3-\lambda)\ldots(\lambda_n-\lambda). \tag{38}$$

Comparing (37) and (38) we have

$$\left.\begin{aligned} \alpha_1 &= \lambda_1+\lambda_2+\ldots+\lambda_n, \\ \alpha_2 &= \lambda_1\lambda_2+\lambda_1\lambda_3+\ldots+\lambda_1\lambda_n+\lambda_2\lambda_3+\lambda_2\lambda_4+ \\ &\qquad +\ldots+\lambda_2\lambda_n+\ldots+\lambda_{n-1}\lambda_n, \\ \alpha_3 &= \lambda_1\lambda_2\lambda_3+\ldots+\lambda_{n-2}\lambda_{n-1}\lambda_n, \\ &\;\;\vdots \\ \alpha_n &= \lambda_1\lambda_2\ldots\lambda_n. \end{aligned}\right\} \tag{39}$$

Two important results now follow.

(a) By putting $\lambda = 0$ in (37) we have

$$\alpha_n = |\mathbf{A}| = \lambda_1\lambda_2\ldots\lambda_n, \tag{40}$$

showing that the product of the n eigenvalues of $\mathbf{A}$ is equal to its determinant. It follows that a matrix is singular if it has a zero eigenvalue, and non-singular if all its eigenvalues are non-zero.

(b) By inspection

$$\alpha_1 = a_{11}+a_{22}+\ldots+a_{nn} = \lambda_1+\lambda_2+\ldots+\lambda_n, \tag{41}$$

using (39). Hence

$$Tr\,\mathbf{A} = \sum_{i=1}^{n} a_{ii} = \sum_{i=1}^{n} \lambda_i. \tag{42}$$

In other words, the sum of the eigenvalues of a matrix is equal to its trace.

These two results may be easily verified for the matrices of Examples 1 and 2. For, from Example 1,

$$\sum_{i=1}^{2} \lambda_i = 7 = 4+3 = Tr\,\mathbf{A}, \tag{43}$$

and

$$\lambda_1 \lambda_2 = 5 \cdot 2 = 10 = |\mathbf{A}|. \tag{44}$$

Likewise for Example 2

$$\sum_{i=1}^{3} \lambda_i = -1 - i + i = -1 = Tr\,\mathbf{A}, \tag{45}$$

and

$$\lambda_1 \lambda_2 \lambda_3 = (-1)(-i)(i) = -1 = |\mathbf{A}|. \tag{46}$$

Some further results may also be proved. Suppose $\mathbf{A}$ is an nth order matrix. Then

$$|\mathbf{A} - \lambda I| = |\tilde{\mathbf{A}} - \lambda I|, \tag{47}$$

so $\mathbf{A}$ and its transpose $\tilde{\mathbf{A}}$ have the same eigenvalues. However, $\mathbf{A}$ and $\tilde{\mathbf{A}}$ will have different eigenvectors unless, of course, $\mathbf{A}$ is symmetric ($\mathbf{A} = \tilde{\mathbf{A}}$). Furthermore, if $\lambda_1, \lambda_2, \ldots, \lambda_n$ are the eigenvalues of $\mathbf{A}$, the matrix $k\mathbf{A}$ where k is an arbitrary scalar has eigenvalues $k\lambda_1, k\lambda_2, \ldots, k\lambda_n$. This follows since

$$|k\mathbf{A} - k\lambda\mathbf{I}| = |k(\mathbf{A} - \lambda\mathbf{I})| = k^n\,|(\mathbf{A} - \lambda\mathbf{I})|. \tag{48}$$

We can also show that the eigenvalues of $\mathbf{A}^{-1}$ are the inverses of the eigenvalues of $\mathbf{A}$, provided none of the eigenvalues of $\mathbf{A}$ are zero. For

$$|\mathbf{A} - \lambda\mathbf{I}| = \left| -\lambda\mathbf{A}\left(\mathbf{A}^{-1} - \frac{\mathbf{I}}{\lambda}\right) \right| = \pm\lambda^n |\mathbf{A}| \left| \mathbf{A}^{-1} - \frac{\mathbf{I}}{\lambda} \right|, \tag{49}$$

which shows that if $\mathbf{A}$ has eigenvalues λ, then $\mathbf{A}^{-1}$ has eigenvalues $1/\lambda$.

5.4 Repeated eigenvalues

In Examples 1 and 2 of 5.2 the eigenvalues of each matrix were all different. We may now prove quite generally that if $\mathbf{X}_1, \mathbf{X}_2, \ldots, \mathbf{X}_n$ are the n eigenvectors corresponding to n different eigenvalues

$\lambda_1, \lambda_2, \ldots, \lambda_n$ then the eigenvectors are linearly independent – that is, there is no linear relationship between them of the type

$$c_1\mathbf{X}_1 + c_2\mathbf{X}_2 + \ldots + c_n\mathbf{X}_n = 0, \tag{50}$$

where the c_i are constants, except when $c_1 = c_2 = \ldots = c_n = 0$. Now, since $\mathbf{AX}_j = \lambda_j\mathbf{X}_j (j = 1, 2, \ldots, n)$, we have

$$(\mathbf{A} - \lambda_i\mathbf{I})\mathbf{X}_j = (\lambda_j - \lambda_i)\mathbf{X}_j. \tag{51}$$

Suppose a linear relation of the type (50) does exist for some non-zero values of c_i. Consider

$$(\mathbf{A} - \lambda_2\mathbf{I})(c_1\mathbf{X}_1 + c_2\mathbf{X}_2 + \ldots + c_n\mathbf{X}_n) = 0. \tag{52}$$

Using (51), (52) becomes

$$c_1(\lambda_1 - \lambda_2)\mathbf{X}_1 + c_3(\lambda_3 - \lambda_2)\mathbf{X}_3 + \ldots + c_n(\lambda_n - \lambda_2)\mathbf{X}_n = 0 \tag{53}$$

in which the $\mathbf{X}_2$ eigenvector is missing. Proceeding in a similar way and operating with $(\mathbf{A} - \lambda_3\mathbf{I})$, $(\mathbf{A} - \lambda_4\mathbf{I}), \ldots, (\mathbf{A} - \lambda_n\mathbf{I})$ we eliminate in turn $\mathbf{X}_3, \mathbf{X}_4, \ldots, \mathbf{X}_n$ and arrive at

$$c_1(\lambda_1 - \lambda_2)(\lambda_1 - \lambda_3)(\lambda_1 - \lambda_4)\ldots(\lambda_1 - \lambda_n)\mathbf{X}_1 = 0. \tag{54}$$

Now by assumption all λ_i are different. Hence, since $\mathbf{X}_1 \neq 0$, $c_1 = 0$. In a similar fashion we can, by operating on (50) by $(\mathbf{A} - \lambda_1\mathbf{I})$, $(\mathbf{A} - \lambda_3\mathbf{I}), \ldots, (\mathbf{A} - \lambda_n\mathbf{I})$ eliminate in turn $\mathbf{X}_1, \mathbf{X}_3, \ldots, \mathbf{X}_n$ and obtain

$$c_2(\lambda_2 - \lambda_1)(\lambda_2 - \lambda_3)\ldots(\lambda_2 - \lambda_n)\mathbf{X}_2 = 0 \tag{55}$$

showing that $c_2 = 0$.

In this way it can be shown that provided all the λ_i are different then $c_1 = c_2 = \ldots = c_n = 0$. Hence no linear relationship exists between the eigenvectors and they are consequently linearly independent.

If, however, two or more eigenvalues are equal then the c_i need not necessarily all be zero, and the eigenvectors may be either linearly dependent or linearly independent (see Examples 3 and 6 which follow and also 5.6 dealing with real symmetric matrices).

Example 3. Consider the matrix

$$\mathbf{A} = \begin{pmatrix} 2 & 1 & 2 \\ 0 & 2 & 3 \\ 0 & 0 & 5 \end{pmatrix}. \tag{56}$$

The characteristic equation is

$$\begin{vmatrix} 2-\lambda & 1 & 2 \\ 0 & 2-\lambda & 3 \\ 0 & 0 & 5-\lambda \end{vmatrix} = 0, \tag{57}$$

which has the solutions

$$\lambda_1 = 5, \qquad \lambda_2 = 2, \qquad \lambda_3 = 2. \tag{58}$$

We now obtain the eigenvectors corresponding to each eigenvalue, noting that two of the eigenvalues are the same.

Case $\lambda_1 = 5$

The homogeneous equations (6) become

$$\left[\begin{pmatrix} 2 & 1 & 2 \\ 0 & 2 & 3 \\ 0 & 0 & 5 \end{pmatrix} - 5\begin{pmatrix} 1 & 0 & 0 \\ 0 & 1 & 0 \\ 0 & 0 & 1 \end{pmatrix}\right]\begin{pmatrix} x_1^{(1)} \\ x_2^{(1)} \\ x_3^{(1)} \end{pmatrix} = \begin{pmatrix} 0 \\ 0 \\ 0 \end{pmatrix}, \tag{59}$$

which give

$$\begin{aligned} -3x_1^{(1)} + x_2^{(1)} + 2x_3^{(1)} &= 0, \\ -x_2^{(1)} + x_3^{(1)} &= 0. \end{aligned} \tag{60}$$

Hence $x_1^{(1)} = x_2^{(1)} = x_3^{(1)}$. Normalising to unit length as before we have

$$x_1^{(1)} = x_2^{(1)} = x_3^{(1)} = \frac{1}{\sqrt{3}} \tag{61}$$

and consequently

$$\mathbf{X}_1 = \begin{pmatrix} \dfrac{1}{\sqrt{3}} \\ \dfrac{1}{\sqrt{3}} \\ \dfrac{1}{\sqrt{3}} \end{pmatrix}. \tag{62}$$

Case $\lambda_2 = \lambda_3 = 2$

Here equations (6) reduce to

$$\begin{pmatrix} 0 & 1 & 2 \\ 0 & 0 & 3 \\ 0 & 0 & 3 \end{pmatrix}\begin{pmatrix} x_1^{(2,3)} \\ x_2^{(2,3)} \\ x_3^{(2,3)} \end{pmatrix} = \begin{pmatrix} 0 \\ 0 \\ 0 \end{pmatrix}, \tag{63}$$

which lead to

$$x_1^{(2,3)} \text{ arbitrary}, \qquad x_2^{(2,3)} = 0, \qquad x_3^{(2,3)} = 0. \tag{64}$$

Consequently, in this case there is only one eigenvector – namely

$$\mathbf{X}_2 = \begin{pmatrix} \alpha \\ 0 \\ 0 \end{pmatrix}, \tag{65}$$

where α is an arbitrary parameter. Normalising $\mathbf{X}_2$ to unit length leads to $\alpha = 1$. The eigenvector $\mathbf{X}_3$ has also to have the same form as $\mathbf{X}_2$ and is consequently linearly dependent on $\mathbf{X}_2$.

In this example the matrix $\mathbf{A}$, in virtue of its repeated eigenvalues, has only two linearly independent eigenvectors.

5.5 Orthogonal properties of eigenvectors

In 5.3 we showed that $\mathbf{A}$ and $\tilde{\mathbf{A}}$ have the same eigenvalues. However, $\mathbf{A}$ and $\tilde{\mathbf{A}}$ will have different eigenvectors (unless $\mathbf{A}$ is symmetric – see 5.6).

Now let $\mathbf{A}$ be a non-symmetric square matrix and consider

$$\mathbf{AX}_i = \lambda_i \mathbf{X}_i, \tag{66}$$

where $\mathbf{X}_i$ is the column eigenvector corresponding to the ith eigenvalue of λ_i. Let $\mathbf{Y}_i$ be the column eigenvector of $\tilde{\mathbf{A}}$ corresponding to the ith eigenvalue λ_i. Then

$$\tilde{\mathbf{A}}\mathbf{Y}_i = \lambda_i \mathbf{Y}_i. \tag{67}$$

Taking the transpose of (67) we have (using (69), Chapter 2)

$$\tilde{\mathbf{Y}}_i \mathbf{A} = \lambda_i \tilde{\mathbf{Y}}_i, \tag{68}$$

which shows that $\tilde{\mathbf{Y}}_i$ is a *row* eigenvector of $\mathbf{A}$.

From (66) we have by premultiplying by $\tilde{\mathbf{Y}}_j$

$$\tilde{\mathbf{Y}}_j \mathbf{AX}_i = \lambda_i \tilde{\mathbf{Y}}_j \mathbf{X}_i, \tag{69}$$

and from (68) by post-multiplying by $\mathbf{X}_j$

$$\tilde{\mathbf{Y}}_i \mathbf{AX}_j = \lambda_i \tilde{\mathbf{Y}}_i \mathbf{X}_j. \tag{70}$$

By writing i for j and j for i in (70) it follows that

$$\tilde{\mathbf{Y}}_j \mathbf{AX}_i = \lambda_j \tilde{\mathbf{Y}}_j \mathbf{X}_i. \tag{71}$$

Comparing (69) and (71) we find

$$(\lambda_j - \lambda_i)\tilde{\mathbf{Y}}_j \mathbf{X}_i = 0. \tag{72}$$

Hence if $\lambda_i \neq \lambda_j$ then

$$\tilde{\mathbf{Y}}_j \mathbf{X}_i = 0, \tag{73}$$

which shows that the row eigenvector $\tilde{\mathbf{Y}}_j$ corresponding to any eigenvalue of a general square matrix is orthogonal to the column eigenvector $\mathbf{X}_i$ corresponding to any different eigenvalue.

Example 4. We take the matrix of Example 1

$$\mathbf{A} = \begin{pmatrix} 4 & 1 \\ 2 & 3 \end{pmatrix} \tag{74}$$

whose eigenvalues are $\lambda_1 = 2, \lambda_2 = 5$. The corresponding normalised column eigenvectors are respectively

$$\mathbf{X}_1 = \begin{pmatrix} -\dfrac{1}{\sqrt{5}} \\ \dfrac{2}{\sqrt{5}} \end{pmatrix}, \qquad \mathbf{X}_2 = \begin{pmatrix} \dfrac{1}{\sqrt{2}} \\ \dfrac{1}{\sqrt{2}} \end{pmatrix}. \tag{75}$$

Now the row eigenvectors $\tilde{\mathbf{Y}}_1$ and $\tilde{\mathbf{Y}}_2$ are obtained in the following way.

Case $\lambda_1 = 2$

Equations (68) become

$$(y_1^{(1)} \quad y_2^{(1)}) \left[\begin{pmatrix} 4 & 1 \\ 2 & 3 \end{pmatrix} - 2\begin{pmatrix} 1 & 0 \\ 0 & 1 \end{pmatrix}\right] = \begin{pmatrix} 0 \\ 0 \end{pmatrix}, \tag{76}$$

or

$$(y_1^{(1)} \quad y_2^{(1)}) \begin{pmatrix} 2 & 1 \\ 2 & 1 \end{pmatrix} = \begin{pmatrix} 0 \\ 0 \end{pmatrix}, \tag{77}$$

where $y^{(1)}$, $y_2^{(1)}$ are the elements of $\tilde{\mathbf{Y}}_1$.

Equations (76) now give

$$y_1^{(1)} + y_2^{(2)} = 0 \quad \text{(twice)}. \tag{78}$$

Hence $y_1^{(1)} = -y_2^{(1)}$, and the normalised row eigenvector

$$\tilde{\mathbf{Y}}_1 = \begin{pmatrix} \dfrac{1}{\sqrt{2}} & -\dfrac{1}{\sqrt{2}} \end{pmatrix}. \tag{79}$$

Case $\lambda_2 = 5$

Here (68) becomes, after simplification,

$$(y_1^{(2)} \quad y_2^{(2)}) \begin{pmatrix} -1 & 1 \\ 2 & -2 \end{pmatrix} = \begin{pmatrix} 0 \\ 0 \end{pmatrix}, \tag{80}$$

which gives

$$y_1^{(2)} = 2y_2^{(2)} \quad \text{(twice)}. \tag{81}$$

Normalising to unit length we have

$$\tilde{\mathbf{Y}}_2 = \begin{pmatrix} \dfrac{2}{\sqrt{5}} & \dfrac{1}{\sqrt{5}} \end{pmatrix}. \tag{82}$$

Now since both eigenvalues are different, the row eigenvector of one eigenvalue should be orthogonal to the column eigenvector of the other eigenvector (see (73)). This is easily verified since (using (75) and (82))

$$\tilde{\mathbf{Y}}_2\mathbf{X}_1 = \begin{pmatrix} \frac{2}{\sqrt{5}} & \frac{1}{\sqrt{5}} \end{pmatrix} \begin{pmatrix} -\frac{1}{\sqrt{5}} \\ \frac{2}{\sqrt{5}} \end{pmatrix} = 0 \tag{83}$$

and

$$\tilde{\mathbf{Y}}_1\mathbf{X}_2 = \begin{pmatrix} \frac{1}{\sqrt{2}} & -\frac{1}{\sqrt{2}} \end{pmatrix} \begin{pmatrix} \frac{1}{\sqrt{2}} \\ \frac{1}{\sqrt{2}} \end{pmatrix} = 0. \tag{84}$$

5.6 Real symmetric matrices

In all the examples given in this chapter so far, the elements of the basic matrix **A** have been real numbers. However, as Example 2 shows, the eigenvalues and eigenvectors of a real matrix may be complex. We now show that provided **A** is real and symmetric the eigenvalues (and consequently the eigenvectors) are necessarily real.

Suppose **A** is a real symmetric matrix of order n. Then

$$\mathbf{AX}_i = \lambda_i\mathbf{X}_i, \tag{85}$$

where $\mathbf{X}_i$ is the eigenvector corresponding to the eigenvalue λ_i. Taking the complex conjugate of (85) we find

$$\mathbf{AX}_i^* = \lambda_i^*\mathbf{X}_i^* \quad (\text{since } \mathbf{A} = \mathbf{A}^*). \tag{86}$$

Hence, using (85) and (86),

$$\widetilde{\mathbf{X}_i^*}\mathbf{AX}_i - \tilde{\mathbf{X}}_i\mathbf{AX}_i^* = (\lambda_i - \lambda_i^*)\tilde{\mathbf{X}}_i\mathbf{X}_i^*, \tag{87}$$

since $(\widetilde{\mathbf{X}_i^*})\mathbf{X}_i = \tilde{\mathbf{X}}_i\mathbf{X}_i^*$.

Now $\tilde{\mathbf{X}}_i\mathbf{AX}_i^*$ is a number and the transpose of a number is itself. Consequently

$$\tilde{\mathbf{X}}_i\mathbf{AX}_i^* = (\widetilde{\tilde{\mathbf{X}}_i\mathbf{AX}_i^*}) = \widetilde{\mathbf{X}_i^*}\tilde{\mathbf{A}}\mathbf{X}_i \tag{88}$$

$$= \widetilde{\mathbf{X}_i^*}\mathbf{AX}_i \quad (\text{since } \mathbf{A} = \tilde{\mathbf{A}}, \mathbf{A} \text{ being symmetric}). \tag{89}$$

Using (89), (87) gives

$$(\lambda_i - \lambda_i^*)\tilde{\mathbf{X}}_i\mathbf{X}_i^* = 0. \tag{90}$$

Since $\tilde{\mathbf{X}}_i \mathbf{X}_i^*$ is just the square of the length of the eigenvector $\mathbf{X}_i$ it is both real and positive, and hence

$$\lambda_i = \lambda_i^* \tag{91}$$

showing that λ_i is a real quantity.

Another important result concerning symmetric matrices can be readily deduced from 5.5 by putting $\tilde{\mathbf{A}} = \mathbf{A}$. It then follows from (67) that $\mathbf{Y}_i$ is an eigenvector of $\mathbf{A}$, and consequently (73) gives

$$\tilde{\mathbf{X}}_j \mathbf{X}_i = 0 \quad \text{for} \quad \lambda_i \neq \lambda_j. \tag{92}$$

In other words, eigenvectors corresponding to different eigenvalues of a symmetric matrix are orthogonal.

Example 5. The real symmetric matrix

$$\mathbf{A} = \begin{pmatrix} 3 & 4 \\ 4 & -3 \end{pmatrix} \tag{93}$$

has eigenvalues 5 and -5 (both real). The normalised eigenvectors appropriate to these two eigenvalues are respectively

$$\mathbf{X}_1 = \begin{pmatrix} \dfrac{2}{\sqrt{5}} \\ \dfrac{1}{\sqrt{5}} \end{pmatrix} \quad \text{and} \quad \mathbf{X}_2 = \begin{pmatrix} -\dfrac{1}{\sqrt{5}} \\ \dfrac{2}{\sqrt{5}} \end{pmatrix}, \tag{94}$$

which are orthogonal since $\tilde{\mathbf{X}}_1 \mathbf{X}_2 = 0$.

The orthogonality property of eigenvectors expressed by (92) is, as we have seen, true if the eigenvalues are different. It can also be proved that if out of the set of n eigenvalues k of them are the same (i.e. a repeated root of the characteristic equation of multiplicity k) then there are k orthogonal eigenvectors corresponding to this particular repeated eigenvalue, and that each of these eigenvectors is orthogonal to the eigenvectors corresponding to the other $n-k$ different eigenvalues.

Example 6. Consider the matrix

$$\mathbf{A} = \begin{pmatrix} 2 & 0 & 1 \\ 0 & 3 & 0 \\ 1 & 0 & 2 \end{pmatrix}. \tag{95}$$

The characteristic equation is

$$\begin{vmatrix} 2-\lambda & 0 & 1 \\ 0 & 3-\lambda & 0 \\ 1 & 0 & 2-\lambda \end{vmatrix} = 0, \tag{96}$$

which has the roots

$$\lambda_1 = 1, \qquad \lambda_2 = \lambda_3 = 3. \tag{97}$$

The root of value 3 is therefore a repeated root of multiplicity two. The eigenvectors are now found in the usual way.

Case $\lambda_1 = 1$

The homogeneous equations (6) become

$$\left[\begin{pmatrix} 2 & 0 & 1 \\ 0 & 3 & 0 \\ 1 & 0 & 2 \end{pmatrix} - 1 \begin{pmatrix} 1 & 0 & 0 \\ 0 & 1 & 0 \\ 0 & 0 & 1 \end{pmatrix}\right] \begin{pmatrix} x_1^{(1)} \\ x_2^{(1)} \\ x_3^{(1)} \end{pmatrix} = \begin{pmatrix} 0 \\ 0 \\ 0 \end{pmatrix}, \tag{98}$$

which give

$$\left.\begin{aligned} x_1^{(1)} + x_3^{(1)} &= 0, \\ 2x_2^{(1)} &= 0, \\ x_1^{(1)} + x_3^{(1)} &= 0. \end{aligned}\right\} \tag{99}$$

Hence $x_1^{(1)} = -x_3^{(1)}$, and $x_2^{(1)} = 0$. Normalising to unit length as before we have

$$x_1^{(1)} = \frac{1}{\sqrt{2}}, \qquad x_2^{(1)} = 0, \qquad x_3^{(1)} = -\frac{1}{\sqrt{2}}, \tag{100}$$

and consequently

$$\mathbf{X}_1 = \begin{pmatrix} \dfrac{1}{\sqrt{2}} \\ 0 \\ -\dfrac{1}{\sqrt{2}} \end{pmatrix}. \tag{101}$$

Case $\lambda_2 = \lambda_3 = 3$

Here the equations (6) reduce to

$$\begin{pmatrix} -1 & 0 & 1 \\ 0 & 0 & 0 \\ 1 & 0 & -1 \end{pmatrix} \begin{pmatrix} x_1^{(2,3)} \\ x_2^{(2,3)} \\ x_3^{(2,3)} \end{pmatrix} = \begin{pmatrix} 0 \\ 0 \\ 0 \end{pmatrix}, \tag{102}$$

which give

$$x_1^{(2,3)} = x_3^{(2,3)}, \qquad x_2^{(2,3)} \text{ arbitrary.} \tag{103}$$

Two mutually orthogonal normalised eigenvectors satisfying (103) are

$$\mathbf{X}_2 = \begin{pmatrix} \frac{1}{\sqrt{2}} \\ 0 \\ \frac{1}{\sqrt{2}} \end{pmatrix} \quad \text{and} \quad \mathbf{X}_3 = \begin{pmatrix} 0 \\ 1 \\ 0 \end{pmatrix}, \tag{104}$$

both of which are orthogonal to the eigenvector $\mathbf{X}_1$ given by (101).

The results of this section (which are of importance in Chapter 6) may be summarised as follows:

(a) The eigenvalues of a real symmetric matrix are real.

(b) For a real symmetric matrix of order n there are n mutually orthogonal (and normalisable) eigenvectors $\mathbf{X}_i$ irrespective of whether the eigenvalues are all different or not. Assuming the $\mathbf{X}_i$ are normalised to unit length this result may be written as

$$\tilde{\mathbf{X}}_i \mathbf{X}_j = \delta_{ij}, \tag{105}$$

where δ_{ij} is the Kronecker delta symbol (see Chapter 2, equation (46)).

Eigenvectors satisfying (105) are called *orthonormal.*

5.7 Hermitian matrices

In Chapter 2, 2.3(k) it was shown that a real symmetric matrix ($\mathbf{A} = \tilde{\mathbf{A}}$) is just the real counterpart of a Hermitian matrix ($\mathbf{A} = \tilde{\mathbf{A}}^*$). Corresponding results to those obtained in the last section apply therefore to Hermitian matrices – namely:

(a) The eigenvalues of a Hermitian matrix are real.

(b) For a Hermitian matrix of order n there are n mutually orthogonal (and normalisable) eigenvectors $\mathbf{X}_i$ irrespective of whether the eigenvalues are all different or not. If the $\mathbf{X}_i$ are normalised to unit length the corresponding result to (105) is

$$\widetilde{\mathbf{X}_i^*} \mathbf{X}_j = \delta_{ij}. \tag{106}$$

Example 7. The matrix

$$\mathbf{A} = \begin{pmatrix} 1 & 1+i \\ 1-i & 2 \end{pmatrix} \tag{107}$$

is Hermitian. Its characteristic equation is

$$\begin{vmatrix} 1-\lambda & 1+i \\ 1-i & 2-\lambda \end{vmatrix} = 0, \tag{108}$$

which gives eigenvalues

$$\lambda_1 = 0, \qquad \lambda_2 = 3 \quad \text{(both real)}. \tag{109}$$

The normalised column eigenvectors in these two cases are respectively

$$\mathbf{X}_1 = \begin{pmatrix} -\dfrac{(1+i)}{\sqrt{3}} \\ \dfrac{1}{\sqrt{3}} \end{pmatrix} \quad \text{and} \quad \mathbf{X}_2 = \begin{pmatrix} \dfrac{1+i}{\sqrt{6}} \\ \sqrt{\dfrac{2}{3}} \end{pmatrix}. \tag{110}$$

Then

$$\widetilde{\mathbf{X}_2^*}\mathbf{X}_1 = \begin{pmatrix} \dfrac{1-i}{\sqrt{6}} & \sqrt{\dfrac{2}{3}} \end{pmatrix} \begin{pmatrix} -\dfrac{(1+i)}{\sqrt{3}} \\ \dfrac{1}{\sqrt{3}} \end{pmatrix} = 0, \tag{111}$$

showing that $\mathbf{X}_1$ and $\mathbf{X}_2$ are orthogonal.

5.8 Non-homogeneous equations

Consider now the non-homogeneous system of equations

$$\mathbf{AX} - \lambda\mathbf{X} = \mathbf{B}, \tag{112}$$

where $\mathbf{A}$ is a real symmetric n^{th} order matrix, $\mathbf{X}$ is a column matrix of order $(n \times 1)$, λ is a *given* constant, and $\mathbf{B}$ a real $(n \times 1)$ column matrix. Now the n eigenvectors of $\mathbf{A}$ are known to form a set of n mutually orthogonal n-dimensional vectors and consequently are linearly independent. Hence any other n-dimensional vector may be expanded as a linear combination of these eigenvectors. Let the normalised eigenvectors of $\mathbf{A}$ be $\mathbf{X}_i^0$ so that

$$\mathbf{AX}_i^0 = \lambda_i \mathbf{X}_i^0. \tag{113}$$

Expressing $\mathbf{X}$ as a linear combination of the $\mathbf{X}_i$ by the relation

$$\mathbf{X} = \sum_{j=1}^{n} \alpha_j \mathbf{X}_j^0, \tag{114}$$

where the α_j are constants, and inserting (114) into (112) we find

$$\mathbf{A}\sum_{j=1}^{n}\alpha_j\mathbf{X}_j^0-\lambda\sum_{j=1}^{n}\alpha_j\mathbf{X}_j^0=\mathbf{B}. \tag{115}$$

Inserting (113) into the first term of (115) gives

$$\sum_{j=1}^{n}\lambda_j\alpha_j\mathbf{X}_j^0-\lambda\sum_{j=1}^{n}\alpha_j\mathbf{X}_j^0=\mathbf{B}. \tag{116}$$

We now pre-multiply (116) through by $\widetilde{\mathbf{X}_k^0}$, where $\mathbf{X}_k^0$ is the kth eigenvector of $\mathbf{A}$. Then

$$\widetilde{\mathbf{X}_k^0}\sum_{j=1}^{n}\lambda_j\alpha_j\mathbf{X}_j^0-\lambda\widetilde{\mathbf{X}_k^0}\sum_{j=1}^{n}\alpha_j\mathbf{X}_j^0=\widetilde{\mathbf{X}_k^0}\mathbf{B}. \tag{117}$$

Now owing to the orthogonality property of the $\mathbf{X}_k^0$ and the assumption that the $\mathbf{X}_k^0$ are normalised to unit length we have $\widetilde{\mathbf{X}_k^0}\mathbf{X}_j^0 = \delta_{kj}$ (see 105)). Consequently (117) becomes

$$\lambda_k\alpha_k-\lambda\alpha_k=\widetilde{\mathbf{X}_k^0}\mathbf{B}, \tag{118}$$

or

$$\alpha_k=\frac{\widetilde{\mathbf{X}_k^0}\mathbf{B}}{\lambda_k-\lambda}, \qquad (k=1,2,\dots,n). \tag{119}$$

Hence, from (114), the solution of the set of equations (112) is

$$\mathbf{X}=\sum_{j=1}^{n}\left(\frac{\widetilde{\mathbf{X}_j^0}\mathbf{B}}{\lambda_j-\lambda}\right)\mathbf{X}_j^0 \tag{120}$$

provided λ is *not* one of the eigenvalues λ_j of $\mathbf{A}$.

Example 8. Consider the set of equations (112) with

$$\mathbf{A}=\begin{pmatrix}1 & 2\\ 2 & 1\end{pmatrix}, \qquad \lambda=1, \qquad \mathbf{B}=\begin{pmatrix}\sqrt{2}\\ 2\sqrt{2}\end{pmatrix}. \tag{121}$$

The eigenvalues of $\mathbf{A}$ are easily found to be

$$\lambda_1=3, \qquad \lambda_2=-1, \tag{122}$$

and the corresponding normalised eigenvectors are

$$\mathbf{X}_1^0=\begin{pmatrix}\frac{1}{\sqrt{2}}\\ \frac{1}{\sqrt{2}}\end{pmatrix}, \qquad \mathbf{X}_2^0=\begin{pmatrix}\frac{1}{\sqrt{2}}\\ -\frac{1}{\sqrt{2}}\end{pmatrix}. \tag{123}$$

Consequently, using (120), the solution **X** of (112) is given by

$$\mathbf{X} = \begin{pmatrix} \frac{1}{\sqrt{2}} & \frac{1}{\sqrt{2}} \end{pmatrix} \begin{pmatrix} \frac{\sqrt{2}}{2\sqrt{2}} \end{pmatrix} \frac{1}{3-1} \begin{pmatrix} \frac{1}{\sqrt{2}} \\ \frac{1}{\sqrt{2}} \end{pmatrix} +$$

$$+ \begin{pmatrix} \frac{1}{\sqrt{2}} & -\frac{1}{\sqrt{2}} \end{pmatrix} \begin{pmatrix} \frac{\sqrt{2}}{2\sqrt{2}} \end{pmatrix} \frac{1}{-1-1} \begin{pmatrix} \frac{1}{\sqrt{2}} \\ -\frac{1}{\sqrt{2}} \end{pmatrix} \tag{124}$$

$$= \frac{3}{2} \begin{pmatrix} \frac{1}{\sqrt{2}} \\ \frac{1}{\sqrt{2}} \end{pmatrix} + \frac{1}{2} \begin{pmatrix} \frac{1}{\sqrt{2}} \\ -\frac{1}{\sqrt{2}} \end{pmatrix} = \begin{pmatrix} \sqrt{2} \\ \frac{1}{\sqrt{2}} \end{pmatrix}. \tag{125}$$

PROBLEMS 5

1. Obtain eigenvalues and eigenvectors normalised to unit length for each of the following matrices:

(a) $\begin{pmatrix} 1 & -8 \\ 2 & 11 \end{pmatrix}$, (b) $\begin{pmatrix} 2 & 0 \\ 1 & 3 \end{pmatrix}$,

(c) $\begin{pmatrix} 1 & 4 & 5 \\ 0 & 2 & 6 \\ 0 & 0 & 3 \end{pmatrix}$, (d) $\begin{pmatrix} 2 & 3 & 1 \\ 0 & 1 & 2 \\ 0 & 0 & 1 \end{pmatrix}$.

2. Obtain eigenvalues and a set of orthonormal eigenvectors for each of the following matrices:

(a) $\begin{pmatrix} 0 & 1 & 0 \\ 1 & 0 & 0 \\ 0 & 0 & 1 \end{pmatrix}$, (b) $\begin{pmatrix} 2 & 2 & 0 \\ 2 & 2 & 0 \\ 0 & 0 & 1 \end{pmatrix}$,

(c) $\begin{pmatrix} 3 & 2 & 2 \\ 2 & 2 & 0 \\ 2 & 0 & 4 \end{pmatrix}$, (d) $\begin{pmatrix} 1 & 1+i \\ 1-i & 2 \end{pmatrix}$.

3. Show that if **A** has eigenvalues $\lambda_1, \lambda_2, \ldots, \lambda_n$ then $\mathbf{A}^m$ (where m is a positive integer) has eigenvalues $\lambda_1^m, \lambda_2^m, \ldots, \lambda_n^m$.

4. Prove that the eigenvalues of a unitary matrix ($\mathbf{A}^{-1} = (\widetilde{\mathbf{A}^*})$) have absolute value 1.

5. Prove that the eigenvalues of a skew-Hermitian matrix ($(\widetilde{\mathbf{A}^*}) = -\mathbf{A}$) are purely imaginary or zero.

6. Show that if $\lambda_1, \lambda_2, \ldots, \lambda_n$ are the eigenvalues of $\mathbf{A}$ then $\lambda_1 - k, \lambda_2 - k, \ldots, \lambda_n - k$ are the eigenvalues of $\mathbf{A} - k\mathbf{I}$.

7. Show that the square matrices $\mathbf{A}$ and $\mathbf{B} = \mathbf{T}^{-1}\mathbf{AT}$ have the same eigenvalues, where $\mathbf{T}$ is an arbitrary non-singular matrix.

8. Prove that if $\mathbf{A}$ and $\mathbf{B}$ are of order n and $\mathbf{A}$ is a non-singular matrix then $\mathbf{A}^{-1}\mathbf{BA}$ and $\mathbf{B}$ have the same eigenvalues.

9. Prove that every eigenvalue of a real orthogonal matrix has absolute value 1. Prove also that both $+1$ and -1 are eigenvalues if the number of rows is even and the determinant has value -1.
   Verify that the matrix

$$\begin{pmatrix} 0 & 0 & 0 & -1 \\ -1 & 0 & 0 & 0 \\ 0 & -1 & 0 & 0 \\ 0 & 0 & -1 & 0 \end{pmatrix}$$

is orthogonal, and determine its eigenvalues.

CHAPTER 6

Diagonalisation of Matrices

6.1 Introduction

In Chapter 5, Example 1, the eigenvalues and eigenvectors of the matrix

$$\mathbf{A} = \begin{pmatrix} 4 & 1 \\ 2 & 3 \end{pmatrix} \tag{1}$$

were found to be respectively $\lambda_1 = 2$, $\lambda_2 = 5$, and

$$\mathbf{X}_1 = \begin{pmatrix} -\dfrac{1}{\sqrt{5}} \\ \dfrac{2}{\sqrt{5}} \end{pmatrix}, \quad \mathbf{X}_2 = \begin{pmatrix} \dfrac{1}{\sqrt{2}} \\ \dfrac{1}{\sqrt{2}} \end{pmatrix}. \tag{2}$$

Consider now the matrix **U**, whose columns are formed from the eigenvectors of **A**. Then

$$\mathbf{U} = \begin{pmatrix} -\dfrac{1}{\sqrt{5}} & \dfrac{1}{\sqrt{2}} \\ \dfrac{2}{\sqrt{5}} & \dfrac{1}{\sqrt{2}} \end{pmatrix} \quad \text{and} \quad \mathbf{U}^{-1} = \begin{pmatrix} -\dfrac{\sqrt{5}}{3} & \dfrac{\sqrt{5}}{3} \\ \dfrac{2\sqrt{2}}{3} & \dfrac{\sqrt{2}}{3} \end{pmatrix}. \tag{3}$$

A straightforward calculation now gives

$$\mathbf{U}^{-1}\mathbf{A}\mathbf{U} = \begin{pmatrix} 2 & 0 \\ 0 & 5 \end{pmatrix} = \mathbf{D} \quad \text{(say)}, \tag{4}$$

showing that the matrix $\mathbf{U}^{-1}\mathbf{A}\mathbf{U}$ is diagonal.

In general, any two matrices **A** and **B** which are related to each other by a relation of the type

$$\mathbf{B} = \mathbf{M}^{-1}\mathbf{A}\mathbf{M} \tag{5}$$

where **M** is *any* non-singular matrix, are said to be similar, equation (5) being called a similarity transformation. The properties of similar matrices are discussed in detail in the next section. For the moment, however, we notice that the matrix **A** of (1) has been transformed into diagonal form (4) by a similarity transformation with **M** equal to **U** (the matrix of the eigenvectors of **A**). It is in fact often possible to carry out the diagonalisation of a matrix in this way, and in the later sections of this chapter we discuss the various conditions under which matrices may be diagonalised. Diagonalisation is an important concept and is useful in many ways. For example, the elements of the kth power (k a positive integer) of a given square matrix **A** are usually difficult to obtain except for small values of k. However, the powers of a diagonal matrix are readily obtained since, if

$$\mathbf{D} = \begin{pmatrix} \alpha & 0 \\ 0 & \beta \end{pmatrix} \tag{6}$$

then

$$\mathbf{D}^k = \begin{pmatrix} \alpha^k & 0 \\ 0 & \beta^k \end{pmatrix}, \tag{7}$$

and similarly for diagonal matrices of higher order. Suppose **A** is an arbitrary matrix of order 2 which is similar to the diagonal matrix **D** of (6). Then

$$\mathbf{D} = \mathbf{M}^{-1}\mathbf{A}\mathbf{M}, \tag{8}$$

where **M** is some non-singular matrix.

Now from (8)

$$\mathbf{D}^2 = (\mathbf{M}^{-1}\mathbf{A}\mathbf{M})(\mathbf{M}^{-1}\mathbf{A}\mathbf{M}) = \mathbf{M}^{-1}\mathbf{A}^2\mathbf{M} \tag{9}$$

since $\mathbf{M}\mathbf{M}^{-1} = \mathbf{I}$.

Similarly

$$\mathbf{D}^3 = \mathbf{M}^{-1}\mathbf{A}^3\mathbf{M} \tag{10}$$

and, in general,

$$\mathbf{D}^k = \mathbf{M}^{-1}\mathbf{A}^k\mathbf{M}. \tag{11}$$

To find $\mathbf{A}^k$ from (11) we now only have to pre-multiply by **M** and post-multiply by $\mathbf{M}^{-1}$ so that

$$\mathbf{M}\mathbf{D}^k\mathbf{M}^{-1} = \mathbf{M}\mathbf{M}^{-1}\mathbf{A}^k\mathbf{M}\mathbf{M}^{-1} = \mathbf{A}^k. \tag{12}$$

Hence, since $\mathbf{D}^k$ may be easily calculated, $\mathbf{A}^k$ may be obtained provided the matrix **M** which diagonalises **A** to **D** is known.

Example 1. To find $\mathbf{A}^8$ given

$$\mathbf{A} = \begin{pmatrix} 4 & 1 \\ 2 & 3 \end{pmatrix}. \tag{13}$$

As already shown (see (4)) **A** is similar to the diagonal matrix

$$\mathbf{D} = \begin{pmatrix} 2 & 0 \\ 0 & 5 \end{pmatrix}. \tag{14}$$

Now

$$\mathbf{D}^8 = \begin{pmatrix} 2 & 0 \\ 0 & 5 \end{pmatrix}^8 = \begin{pmatrix} 2^8 & 0 \\ 0 & 5^8 \end{pmatrix} = \begin{pmatrix} 256 & 0 \\ 0 & 390625 \end{pmatrix}. \tag{15}$$

The matrix **M** which diagonalises **A** is the matrix **U** given by (3). Hence using (12)

$$\mathbf{A}^8 = \mathbf{U}\mathbf{D}^8\mathbf{U}^{-1}$$

$$= \begin{pmatrix} -\dfrac{1}{\sqrt{5}} & \dfrac{1}{\sqrt{2}} \\ \dfrac{2}{\sqrt{5}} & \dfrac{1}{\sqrt{2}} \end{pmatrix} \begin{pmatrix} 256 & 0 \\ 0 & 390625 \end{pmatrix} \begin{pmatrix} -\dfrac{\sqrt{5}}{3} & \dfrac{\sqrt{5}}{3} \\ \dfrac{2\sqrt{2}}{3} & \dfrac{\sqrt{2}}{3} \end{pmatrix}$$

$$= \begin{pmatrix} 260502 & 130123 \\ 260246 & 130379 \end{pmatrix}. \tag{16}$$

Apart from the ease of calculating powers of matrices by first diagonalising them (if possible), another motive can be seen in diagonalisation. Consider as an illustration the problem of the maximisation or minimisation of a function of two variables, say $f(x, y)$. Suppose $f(x, y)$ has a convergent Taylor expansion within some domain of the xy-plane so that if (x_0, y_0) is a typical point in this region then

$$f(x,y) = f(x_0, y_0) + (x - x_0)\left(\frac{\partial f}{\partial x}\right)_{\substack{x=x_0\\y=y_0}} + (y - y_0)\left(\frac{\partial f}{\partial y}\right)_{\substack{x=x_0\\y=y_0}} +$$

$$+ \frac{(x-x_0)^2}{2!}\left(\frac{\partial^2 f}{\partial x^2}\right)_{\substack{x=x_0\\y=y_0}} + (x-x_0)(y-y_0)\left(\frac{\partial^2 f}{\partial x \partial y}\right)_{\substack{x=x_0\\y=y_0}} +$$

$$+ \frac{(y-y_0)^2}{2!}\left(\frac{\partial^2 f}{\partial y^2}\right)_{\substack{x=x_0\\y=y_0}} + \ldots \tag{17}$$

Now the necessary conditions for a stationary value of $f(x, y)$ at (x_0, y_0) are

$$\left(\frac{\partial f}{\partial x}\right)_{\substack{x=x_0\\y=y_0}} = 0, \qquad \left(\frac{\partial f}{\partial y}\right)_{\substack{x=x_0\\y=y_0}} = 0, \tag{18}$$

which are equations determining the values of x_0 and y_0. However, these conditions do not tell us whether the function takes on a maximum value or minimum value at (x_0, y_0) or whether the stationary point is a saddle point. Clearly the nature of the stationary point depends on the quadratic terms in (17) since, using (18),

$$f(x, y) = f(x_0, y_0) + \tfrac{1}{2}Q(x, y), \tag{19}$$

where

$$Q(x, y) = a(x-x_0)^2 + 2h(x-x_0)(y-y_0) + b(y-y_0)^2 \tag{20}$$

and

$$a = \left(\frac{\partial^2 f}{\partial x^2}\right)_{\substack{x=x_0\\y=y_0}}, \qquad h = \left(\frac{\partial^2 f}{\partial x \partial y}\right)_{\substack{x=x_0\\y=y_0}}, \qquad b = \left(\frac{\partial^2 f}{\partial y^2}\right)_{\substack{x=x_0\\y=y_0}}. \tag{21}$$

Writing $x-x_0 = u$, $y-y_0 = v$ we have

$$Q(u, v) = au^2 + 2huv + bv^2. \tag{22}$$

This expression is called a quadratic form in u and v since each term is homogeneous of degree 2. If $Q(u, v) > 0$ for all values of u and v close to zero (i.e. for (x, y) values in the neighbourhood of (x_0, y_0)) then $f(x, y) > f(x_0, y_0)$ and the stationary point corresponds to a minimum of $f(x, y)$. Likewise if $Q(u, v) < 0$ then $f(x, y) < f(x_0, y_0)$ and the stationary point corresponds to a maximum. (The case in which $Q(u, v)$ takes both positive and negative values leads to a saddle point.) Clearly determining the nature of a stationary point is closely related to determining whether a quadratic form is positive definite (i.e. $Q(u, v) > 0$ for non-zero u, v), negative definite (i.e. $Q(u, v) < 0$ for non-zero u, v) or indefinite ($Q(u, v)$ taking both positive and negative values for different u, v values). This can best be done by writing (22) in matrix form as

$$Q = \tilde{\mathbf{S}}\mathbf{A}\mathbf{S}, \tag{23}$$

where

$$\mathbf{S} = \begin{pmatrix} u \\ v \end{pmatrix} \quad \text{and} \quad \mathbf{A} = \begin{pmatrix} a & h \\ h & b \end{pmatrix}. \tag{24}$$

If now Q can be put into diagonal form such that

$$Q = \lambda U^2 + \mu V^2, \tag{25}$$

where λ, μ are constants and U, V are new variables related to u, v, then Q is positive when both λ and μ are positive, and negative when both λ and μ are negative. As we shall see in 6.8 the reduction of a quadratic form to a diagonal form is closely related to the diagonalisation of the matrix **A** associated with the form (see, for example, (23) and (24)).

Now the nature of a quadratic form in two variables such as (22) may easily be determined without using matrix ideas. For from first principles $Q(u, v) > 0$ if $a > 0$ and $ab - h^2 > 0$, and $Q(u, v) < 0$ if $a < 0$, $ab - h^2 > 0$. However, when dealing with functions of many variables say $f(x_1, x_2, \ldots, x_n)$ the nature of the stationary point depends on the nature of the general quadratic form

$$\sum_{i=1}^{n} \sum_{k=1}^{n} a_{ik} u_i u_k = a_{11} u_1^2 + a_{12} u_1 u_2 \pm \ldots + a_{nn} u_n^2 \quad (26)$$

and the matrix approach becomes of prime importance. We shall return to this problem in 6.8.

6.2 Similar matrices

As we have already seen two matrices **A** and **B** are said to be similar if there exists a non-singular matrix **M** such that

$$\mathbf{B} = \mathbf{M}^{-1}\mathbf{A}\mathbf{M}. \quad (27)$$

We now show that similar matrices have the same eigenvalues. For

$$\mathbf{B} - \lambda\mathbf{I} = \mathbf{M}^{-1}\mathbf{A}\mathbf{M} - \lambda\mathbf{I} = \mathbf{M}^{-1}\mathbf{A}\mathbf{M} - \lambda\mathbf{M}^{-1}\mathbf{M} = \mathbf{M}^{-1}(\mathbf{A} - \lambda\mathbf{I})\mathbf{M}, \quad (28)$$

and hence

$$|\mathbf{B} - \lambda\mathbf{I}| = |\mathbf{M}^{-1}(\mathbf{A} - \lambda\mathbf{I})\mathbf{M}| = |\mathbf{M}^{-1}|\,|\mathbf{A} - \lambda\mathbf{I}|\,|\mathbf{M}| = |\mathbf{A} - \lambda\mathbf{I}|, \quad (29)$$

since $|\mathbf{M}^{-1}|\,|\mathbf{M}| = 1$. Consequently the characteristic polynomials of **A** and **B** are identical and so therefore are their eigenvalues.

Suppose now that $\mathbf{X}_i$ is an eigenvector of **A** and $\mathbf{Y}_i$ is an eigenvector of **B** both corresponding to the ith eigenvalue λ_i. Then

$$\mathbf{A}\mathbf{X}_i = \lambda_i \mathbf{X}_i \quad (30)$$

and

$$\mathbf{B}\mathbf{Y}_i = \lambda_i \mathbf{Y}_i \quad (31)$$

But since $\mathbf{B} = \mathbf{M}^{-1}\mathbf{A}\mathbf{M}$ then

$$\mathbf{M}\mathbf{B} = \mathbf{A}\mathbf{M}, \quad (32)$$

and hence, using (31),

$$\mathbf{M}\mathbf{B}\mathbf{Y}_i = \mathbf{A}\mathbf{M}\mathbf{Y}_i = \lambda_i \mathbf{M}\mathbf{Y}_i. \quad (33)$$

Comparing (33) and (30) we see that

$$\mathbf{X}_i = \mathbf{M}\mathbf{Y}_i. \quad (34)$$

Another result follows from the fact that **A** and **B** both have the same characteristic equation for then

$$Tr\,\mathbf{A} = Tr\,\mathbf{B}. \tag{35}$$

Furthermore

$$|\mathbf{B}| = |\mathbf{M}^{-1}\mathbf{A}\mathbf{M}| = |\mathbf{M}^{-1}|\,|\mathbf{A}|\,|\mathbf{M}| = |\mathbf{A}| \tag{36}$$

so that the determinants of similar matrices are equal. To illustrate these results we again return to the matrix

$$\mathbf{A} = \begin{pmatrix} 4 & 1 \\ 2 & 3 \end{pmatrix}. \tag{37}$$

In 6.1 it was found that **A** is similar to the diagonal matrix

$$\mathbf{D} = \begin{pmatrix} 2 & 0 \\ 0 & 5 \end{pmatrix}. \tag{38}$$

Clearly $Tr\,\mathbf{A} = 4+3 = 5+2 = Tr\,\mathbf{D}$, and $|\mathbf{A}| = 10 = |\mathbf{D}|$. As a further example we take

$$\mathbf{A} = \begin{pmatrix} 1 & 3 \\ 2 & 2 \end{pmatrix}, \qquad \mathbf{M} = \begin{pmatrix} 1 & 0 \\ 1 & 1 \end{pmatrix}. \tag{39}$$

Then

$$\mathbf{B} = \mathbf{M}^{-1}\mathbf{A}\mathbf{M} = \begin{pmatrix} 4 & 3 \\ 0 & -1 \end{pmatrix}. \tag{40}$$

Again it is easily verified that $Tr\,\mathbf{B} = Tr\,\mathbf{A}$, and that $|\mathbf{A}| = |\mathbf{B}|$.

6.3 Diagonalisation of a matrix whose eigenvalues are all different

In Chapter 5, 5.4 we showed that if the eigenvalues of a general n^{th} order matrix **A**, say, are all different then a set of n linearly independent eigenvectors always exists. Now let **U** be the square matrix whose columns are the eigenvectors of **A**. Then, if $\mathbf{X}_i$ is the eigenvector corresponding to the ith eigenvalue λ_i, we have

$$\mathbf{U} = (\mathbf{X}_1\mathbf{X}_2\mathbf{X}_3 \dots \mathbf{X}_n), \tag{41}$$

which in terms of the components of the eigenvectors is

$$\mathbf{U} = \begin{pmatrix} x_1^{(1)} & x_1^{(2)} & \dots & x_1^{(n)} \\ x_2^{(1)} & x_2^{(2)} & \dots & x_2^{(n)} \\ \cdot & \cdot & & \cdot \\ \cdot & \cdot & & \cdot \\ \cdot & \cdot & & \cdot \\ x_n^{(1)} & x_n^{(2)} & \dots & x_n^{(n)} \end{pmatrix}. \tag{42}$$

We further write

$$\mathbf{D} = \begin{pmatrix} \lambda_1 & 0 & 0 & . & . & . & 0 \\ 0 & \lambda_2 & 0 & . & . & . & 0 \\ 0 & 0 & \lambda_3 & . & . & . & 0 \\ . & . & & & & & . \\ . & . & & & & & . \\ . & . & & & & & . \\ 0 & 0 & . & . & . & . & \lambda_n \end{pmatrix}. \tag{43}$$

Consequently

$$\mathbf{UD} = \begin{pmatrix} \lambda_1 x_1^{(1)} & \lambda_2 x_1^{(2)} . & . & . & \lambda_n x_1^{(n)} \\ \lambda_1 x_2^{(1)} & \lambda_2 x_2^{(2)} . & . & . & \lambda_n x_2^{(n)} \\ . & . & & & . \\ . & . & & & . \\ . & . & & & . \\ \lambda_1 x_n^{(1)} & \lambda_2 x_n^{(2)} . & . & . & \lambda_n x_n^{(n)} \end{pmatrix} \tag{44}$$

which, in the abbreviated notation of (41), is the matrix

$$(\lambda_1 \mathbf{X}_1 \quad \lambda_2 \mathbf{X}_2 \ldots \lambda_n \mathbf{X}_n). \tag{45}$$

Likewise **AU** is an nth order matrix given by

$$\mathbf{AU} = (\mathbf{AX}_1 \quad \mathbf{AX}_2 \ldots \mathbf{AX}_n). \tag{46}$$

But since $\mathbf{AX}_i = \lambda_i \mathbf{X}_i$ we have

$$\mathbf{AU} = (\lambda_1 \mathbf{X}_1 \quad \lambda_2 \mathbf{X}_2 \ldots \lambda_n \mathbf{X}_n). \tag{47}$$

Hence

$$\mathbf{AU} = \mathbf{UD} \tag{48}$$

or

$$\mathbf{D} = \mathbf{U}^{-1}\mathbf{AU}, \tag{49}$$

where **D** is a diagonal matrix whose elements are the eigenvalues of **A**. (We note here that since all the eigenvalues are assumed different the columns of **U** are linearly independent (see 5.4) and hence $\mathbf{U}^{-1}$ exists. If some of the eigenvalues of **A** were the same the eigenvectors would not necessarily be independent. Consequently two or more columns of $|\mathbf{U}|$ would be proportional giving $|\mathbf{U}| = 0$. Hence $\mathbf{U}^{-1}$ would not exist and diagonalisation could not be carried out.) The general result of this section is as follows:

A matrix **A** with all different eigenvalues may be diagonalised by a similarity transformation $\mathbf{D} = \mathbf{U}^{-1}\mathbf{AU}$, where **U** is the matrix whose columns are the eigenvectors of **A**. The diagonal matrix **D** has as its elements the eigenvalues of **A**.

The origin of the diagonalisation of the matrix of equation (1) is now clear, equation (4) showing that the elements of **D** are indeed the eigenvalues of **A**. As a further illustration of the method of diagonalisation we consider the following example.

Example 2. The matrix of Chapter 5, Example 2,

$$\mathbf{A} = \begin{pmatrix} 1 & -1 & -1 \\ 1 & -1 & 0 \\ 1 & 0 & -1 \end{pmatrix} \tag{50}$$

has eigenvalues $\lambda_1 = -1$, $\lambda_2 = i$, $\lambda_3 = -i$ and corresponding eigenvectors

$$\mathbf{X}_1 = \begin{pmatrix} 0 \\ \dfrac{1}{\sqrt{2}} \\ -\dfrac{1}{\sqrt{2}} \end{pmatrix}, \quad \mathbf{X}_2 = \begin{pmatrix} \dfrac{1+i}{2} \\ \frac{1}{2} \\ \frac{1}{2} \end{pmatrix}, \quad \mathbf{X}_3 = \begin{pmatrix} \dfrac{1-i}{2} \\ \frac{1}{2} \\ \frac{1}{2} \end{pmatrix}. \tag{51}$$

Hence

$$\mathbf{U} = \begin{pmatrix} 0 & \dfrac{1+i}{2} & \dfrac{1-i}{2} \\ \dfrac{1}{\sqrt{2}} & \frac{1}{2} & \frac{1}{2} \\ -\dfrac{1}{\sqrt{2}} & \frac{1}{2} & \frac{1}{2} \end{pmatrix}, \tag{52}$$

$|\mathbf{U}| = -i/\sqrt{2}$, and

$$\mathbf{U}^{-1} = \begin{pmatrix} 0 & \dfrac{1}{\sqrt{2}} & -\dfrac{1}{\sqrt{2}} \\ -i & \dfrac{1+i}{2} & \dfrac{1+i}{2} \\ i & \dfrac{1-i}{2} & \dfrac{1-i}{2} \end{pmatrix}. \tag{53}$$

Hence

$$\mathbf{U}^{-1}\mathbf{A}\mathbf{U} = \begin{pmatrix} -1 & 0 & 0 \\ 0 & i & 0 \\ 0 & 0 & -i \end{pmatrix}. \tag{54}$$

6.4 Matrices with repeated eigenvalues

As shown in Chapter 5, 5.4 the eigenvectors of a matrix with repeated eigenvalues may not be linearly independent. The remarks of the last section then lead us to the fact that such a matrix cannot be diagonalised by a similarity transformation. Consider as an example the matrix of Chapter 5, Example 3, where

$$\mathbf{A} = \begin{pmatrix} 2 & 1 & 2 \\ 0 & 2 & 3 \\ 0 & 0 & 5 \end{pmatrix}, \qquad \lambda_1 = 5, \qquad \lambda_2 = \lambda_3 = 2. \tag{55}$$

Using the results of (62) and (65) (Chapter 5) we have

$$\mathbf{U} = \begin{pmatrix} \dfrac{1}{\sqrt{3}} & 1 & 1 \\ \dfrac{1}{\sqrt{3}} & 0 & 0 \\ \dfrac{1}{\sqrt{3}} & 0 & 0 \end{pmatrix}. \tag{56}$$

Hence $|\mathbf{U}| = 0$, and therefore $\mathbf{U}^{-1}$ does not exist. Consequently $\mathbf{A}$ is not diagonalisable.

In general, a non-symmetric matrix with repeated eigenvalues (such as (55)) is not diagonalisable, but may be reduced to the Jordan normal form. This is a matrix with elements in the leading diagonal, elements equal to 0 or 1 in the next line parallel to and above the leading diagonal, and zeros everywhere else. However, we will not prove this result here.

6.5 Diagonalisation of symmetric matrices

In Chapter 5, 5.6 we have seen that corresponding to any n^{th} order real symmetric matrix $\mathbf{A}$ there are (even if some eigenvalues are repeated) n orthonormal eigenvectors $\mathbf{X}_i$ satisfying the relation

$$\tilde{\mathbf{X}}_i \mathbf{X}_j = \delta_{ij} \tag{57}$$

(see Chapter 5, equation (105)).

We now see that in virtue of (57) the nth order matrix of the eigenvectors $\mathbf{X}_i$ of $\mathbf{A}$, namely,

$$\mathbf{U} = (\mathbf{X}_1 \mathbf{X}_2 \ldots \mathbf{X}_n) \tag{58}$$

satisfies the relation

$$\tilde{\mathbf{U}}\mathbf{U} = \mathbf{I} \tag{59}$$

from which

$$\tilde{\mathbf{U}} = \mathbf{U}^{-1}. \tag{60}$$

Hence **U** is an orthogonal matrix (see Chapter 3, 3.5). Consequently it follows from (49) and (60) that **A** is diagonalised to **D** by an orthogonal matrix **U**, where

$$\mathbf{D} = \mathbf{U}^{-1}\mathbf{A}\mathbf{U} = \tilde{\mathbf{U}}\mathbf{A}\mathbf{U}. \tag{61}$$

The general result obtained here may be stated as follows:

A real symmetric matrix **A** (with distinct or repeated eigenvalues) may be diagonalised by an orthogonal transformation $\mathbf{D} = \tilde{\mathbf{U}}\mathbf{A}\mathbf{U}$, where **U** is the orthogonal matrix whose columns are formed from a set of orthonormal eigenvectors of **A**. The diagonal matrix **D** has as its elements the eigenvalues of **A**.

Example 3. The matrix

$$\mathbf{A} = \begin{pmatrix} 3 & 4 \\ 4 & -3 \end{pmatrix} \tag{62}$$

of Chapter 5, Example 5, has eigenvalues $\lambda_1 = 5$, $\lambda_2 = -5$ and orthonormal eigenvectors

$$\mathbf{X}_1 = \begin{pmatrix} \frac{2}{\sqrt{5}} \\ \frac{1}{\sqrt{5}} \end{pmatrix}, \qquad \mathbf{X}_2 = \begin{pmatrix} -\frac{1}{\sqrt{5}} \\ \frac{2}{\sqrt{5}} \end{pmatrix}. \tag{63}$$

Hence

$$\mathbf{U} = \begin{pmatrix} \frac{2}{\sqrt{5}} & -\frac{1}{\sqrt{5}} \\ \frac{1}{\sqrt{5}} & \frac{2}{\sqrt{5}} \end{pmatrix} \text{ (which is orthogonal)} \tag{64}$$

and

$$\tilde{\mathbf{U}}\mathbf{A}\mathbf{U} = \begin{pmatrix} 5 & 0 \\ 0 & -5 \end{pmatrix} = \mathbf{D}. \tag{65}$$

Example 4. The matrix

$$\mathbf{A} = \begin{pmatrix} 2 & 0 & 1 \\ 0 & 3 & 0 \\ 1 & 0 & 2 \end{pmatrix} \tag{66}$$

of Chapter 5, Example 6, has eigenvalues $\lambda_1 = 1$, $\lambda_2 = \lambda_3 = 3$. An orthonormal set of eigenvectors corresponding to these eigenvalues is (see (101) and (104), Chapter 5)

$$\mathbf{X}_1 = \begin{pmatrix} \frac{1}{\sqrt{2}} \\ 0 \\ -\frac{1}{\sqrt{2}} \end{pmatrix}, \quad \mathbf{X}_2 = \begin{pmatrix} \frac{1}{\sqrt{2}} \\ 0 \\ \frac{1}{\sqrt{2}} \end{pmatrix}, \quad \mathbf{X}_3 = \begin{pmatrix} 0 \\ 1 \\ 0 \end{pmatrix}. \tag{67}$$

Hence

$$\mathbf{U} = \begin{pmatrix} \frac{1}{\sqrt{2}} & \frac{1}{\sqrt{2}} & 0 \\ 0 & 0 & 1 \\ -\frac{1}{\sqrt{2}} & \frac{1}{\sqrt{2}} & 0 \end{pmatrix} \tag{68}$$

and

$$\tilde{\mathbf{U}}\mathbf{A}\mathbf{U} = \begin{pmatrix} 1 & 0 & 0 \\ 0 & 3 & 0 \\ 0 & 0 & 3 \end{pmatrix}. \tag{69}$$

6.6 Diagonalisation of Hermitian matrices

It was found in Chapter 5, equation (106) that the normalised eigenvectors $\mathbf{X}_i$ of a Hermitian matrix satisfy

$$\widetilde{\mathbf{X}_i^*}\mathbf{X}_j = \delta_{ij}. \tag{70}$$

Accordingly the matrix of the normalised eigenvectors

$$\mathbf{U} = (\mathbf{X}_1\mathbf{X}_2 \ldots \mathbf{X}_n) \tag{71}$$

satisfies

$$\widetilde{\mathbf{U}^*}\mathbf{U} = \mathbf{I} \quad (\text{or } \widetilde{\mathbf{U}^*} = \mathbf{U}^{-1}). \tag{72}$$

$\mathbf{U}$ is therefore a unitary matrix. Hence a Hermitian matrix $\mathbf{A}$ can be diagonalised by the unitary matrix $\mathbf{U}$ formed from an orthonormal set of its eigenvectors. For, using (49) and (72), we have

$$\mathbf{U}^{-1}\mathbf{A}\mathbf{U} = \widetilde{\mathbf{U}^*}\mathbf{A}\mathbf{U} = \mathbf{D}, \tag{73}$$

where $\mathbf{D}$ is a diagonal matrix with the eigenvalues of $\mathbf{A}$ as elements.

Example 5. The matrix

$$\mathbf{A} = \begin{pmatrix} 1 & 1+i \\ 1-i & 2 \end{pmatrix} \tag{74}$$

of Chapter 5, Example 7, is Hermitian and has orthonormal eigenvectors

$$\mathbf{X}_1 = \begin{pmatrix} -\dfrac{(1+i)}{\sqrt{3}} \\ \dfrac{1}{\sqrt{3}} \end{pmatrix}, \qquad \mathbf{X}_2 = \begin{pmatrix} \dfrac{1+i}{\sqrt{6}} \\ \sqrt{\dfrac{2}{3}} \end{pmatrix} \tag{75}$$

satisfying (70).

Hence

$$\mathbf{U} = \begin{pmatrix} -\dfrac{(1+i)}{\sqrt{3}} & \dfrac{1+i}{\sqrt{6}} \\ \dfrac{1}{\sqrt{3}} & \sqrt{\dfrac{2}{3}} \end{pmatrix} \tag{76}$$

and

$$\widetilde{\mathbf{U}^*} = \begin{pmatrix} -\dfrac{(1-i)}{\sqrt{3}} & \dfrac{1}{\sqrt{3}} \\ \dfrac{1-i}{\sqrt{6}} & \sqrt{\dfrac{2}{3}} \end{pmatrix}. \tag{77}$$

Consequently

$$\widetilde{\mathbf{U}^*}\mathbf{A}\mathbf{U} = \begin{pmatrix} 0 & 0 \\ 0 & 3 \end{pmatrix} = \mathbf{D}. \tag{78}$$

It is easily verified that 0 and 3 are, in fact, the eigenvalues of **A**.

6.7 Bilinear and quadratic forms

An expression of the type

$$B = \sum_{i=1}^{m} \sum_{j=1}^{n} a_{ij} x_i y_j \tag{79}$$

which is linear and homogeneous in each of the sets of variables $x_1, x_2, \ldots, x_m; y_1, y_2, \ldots, y_n$ is called a bilinear form. For the moment we shall deal only with real forms for which the coefficients a_{ij}, and the variables x_i, y_j are real quantities. Now (79) may be written in terms of matrices as

$$B = \tilde{\mathbf{X}}\mathbf{A}\mathbf{Y}, \tag{80}$$

where

$$\mathbf{X} = \begin{pmatrix} x_1 \\ x_2 \\ \cdot \\ \cdot \\ \cdot \\ x_m \end{pmatrix}, \quad \mathbf{A} = \begin{pmatrix} a_{11} & a_{12} & \cdot & \cdot & \cdot & a_{1n} \\ a_{21} & a_{22} & \cdot & \cdot & \cdot & a_{2n} \\ \cdot & & & & & \cdot \\ \cdot & & & & & \cdot \\ \cdot & & & & & \cdot \\ a_{m1} & \cdot & \cdot & \cdot & \cdot & a_{mn} \end{pmatrix}, \quad \mathbf{Y} = \begin{pmatrix} y_1 \\ y_2 \\ \cdot \\ \cdot \\ \cdot \\ y_n \end{pmatrix}, \tag{81}$$

A being called the matrix of the form.

For example, the bilinear form

$$B_1 = x_1 y_1 + 2y_2 x_1 + 3x_2 y_1 + 4x_2 y_2 + 5y_1 x_3 + 6y_2 x_3 \tag{82}$$

may be written as

$$(x_1 \quad x_2 \quad x_3) \begin{pmatrix} 1 & 2 \\ 3 & 4 \\ 5 & 6 \end{pmatrix} \begin{pmatrix} y_1 \\ y_2 \end{pmatrix}. \tag{83}$$

Likewise

$$B_2 = 6x_1 y_1 + 2x_1 y_2 + 3x_2 y_1 - 4x_2 y_2 \tag{84}$$

$$= (x_1 \quad x_2) \begin{pmatrix} 6 & 2 \\ 3 & -4 \end{pmatrix} \begin{pmatrix} y_1 \\ y_2 \end{pmatrix}. \tag{85}$$

A special case of (80) occurs when **X** and **Y** each have the same number of elements and **A** = **I** (the unit matrix). Then

$$B = \tilde{\mathbf{X}}\mathbf{Y} = x_1 y_1 + x_2 y_2 + \ldots + x_n y_n, \tag{86}$$

which is the matrix form of the scalar product of the two vectors **X** and **Y**.

Bilinear forms lead naturally into quadratic forms when **Y** = **X**, for then

$$B = Q = \sum_{i=1}^{n} \sum_{j=1}^{n} a_{ij} x_i x_j, \tag{87}$$

which is a homogeneous polynomial of degree two in the variables x_i.

In matrix form (87) becomes

$$Q = \tilde{\mathbf{X}}\mathbf{A}\mathbf{X}, \tag{88}$$

where **A** is the matrix of the quadratic form. Now expanding (87) we find

$$Q = a_{11} x_1^2 + (a_{12} + a_{21}) x_1 x_2 + (a_{13} + a_{31}) x_1 x_3 + \ldots + \\ + a_{22} x_2^2 + (a_{23} + a_{32}) x_2 x_3 + \ldots + a_{nn} x_n^2. \tag{89}$$

Writing

$$c_{ik} = \frac{a_{ik}+a_{ki}}{2} \quad \text{for all } i \text{ and } k \tag{90}$$

so that $c_{ik} = c_{ki}$, Q becomes

$$\begin{aligned} c_{11}x_1^2+2c_{12}x_1x_2+2c_{13}x_1x_3+\ldots+ \\ +c_{22}x_2^2+2c_{23}x_2x_3+\ldots+c_{nn}x_n^2, \end{aligned} \tag{91}$$

which in matrix form is

$$\tilde{\mathbf{X}}\mathbf{C}\mathbf{X}, \tag{92}$$

where **C** is a *symmetric* matrix.

For example, the matrix

$$\mathbf{A} = \begin{pmatrix} 1 & -3 \\ 0 & 5 \end{pmatrix} \tag{93}$$

associated with the quadratic form

$$\tilde{\mathbf{X}}\mathbf{A}\mathbf{X} = x_1^2-3x_1x_2+5x_2^2 \tag{94}$$

is non-symmetric. However, writing (94) as

$$x_1^2-\tfrac{3}{2}x_1x_2-\tfrac{3}{2}x_2x_1+5x_2^2 = \tilde{\mathbf{X}}\mathbf{C}\mathbf{X}, \tag{95}$$

we see that the associated symmetric matrix **C** is

$$\begin{pmatrix} 1 & -\frac{3}{2} \\ -\frac{3}{2} & 5 \end{pmatrix}. \tag{96}$$

6.8 Lagrange's reduction of a quadratic form

A real quadratic form

$$Q = \sum_{i=1}^{n}\sum_{j=1}^{n} a_{ij}x_ix_j \tag{97}$$

can be reduced by a variety of methods to the form

$$\alpha_1y_1^2+\alpha_2y_2^2+\ldots+\alpha_ny_n^2, \tag{98}$$

where the y_i are linearly related to the x_i and the α_i are constants. This process is called reducing the quadratic form to a diagonal form – or, more briefly, diagonalisation. One method of diagonalisation (due to Lagrange) consists of continually completing the square, as shown by the following example.

Example 6. Consider

$$= x_1^2+4x_2^2+5x_3^2-2x_2x_3-x_1x_2. \tag{99}$$

Then regrouping the terms in (99) gives

$$= \left(x_1 - \frac{x_2}{2}\right)^2 + \tfrac{15}{4}(x_2 - \tfrac{4}{15}x_3)^2 + 5x_3^2 - \tfrac{4}{15}x_3^2 \tag{100}$$

$$= y_1^2 + \tfrac{15}{4}y_2^2 + \tfrac{71}{15}y_3^2, \tag{101}$$

where

$$\left.\begin{aligned} y_1 &= x_1 - \frac{x_2}{2}, \\ y_2 &= \quad\quad x_2 - \tfrac{4}{15}x_3, \\ y_3 &= \quad\quad\quad\quad x_3. \end{aligned}\right\} \tag{102}$$

6.9 Matrix diagonalisation of a real quadratic form

We now consider the matrix form of Q – namely

$$\tilde{\mathbf{X}}\mathbf{C}\mathbf{X} \quad \text{(see (92))}, \tag{103}$$

where $\mathbf{C}$ is a real symmetric matrix. Suppose we now allow a real non-singular linear transformation of the variables x_i to a new set of variables y_i defined by

$$\mathbf{X} = \mathbf{UY}, \tag{104}$$

where $\mathbf{U}$ is some real non-singular matrix.

Then

$$Q = \widetilde{(\mathbf{UY})}\mathbf{CUY} = \tilde{\mathbf{Y}}\tilde{\mathbf{U}}\mathbf{CUY}. \tag{105}$$

Clearly if $\mathbf{U}$ can be chosen such that $\tilde{\mathbf{U}}\mathbf{CU}$ is a diagonal matrix then Q will be transformed into the diagonal form (98). Although there is frequently no unique way of doing this, an important method already discussed in 6.5 is based on choosing $\mathbf{U}$ to be the matrix of a set of orthonormal eigenvectors of $\mathbf{C}$. The matrix $\mathbf{U}$ is then orthogonal ($\tilde{\mathbf{U}} = \mathbf{U}^{-1}$) and $\tilde{\mathbf{U}}\mathbf{CU}$ is a diagonal matrix $\mathbf{D}$, whose elements are the eigenvalues $\lambda_1, \lambda_2, \ldots, \lambda_n$ of $\mathbf{C}$. Hence, with this particular choice of $\mathbf{U}$, we have

$$Q = \tilde{\mathbf{Y}}\mathbf{D}\mathbf{Y} = (y_1 \quad y_2 \ldots y_n) \begin{pmatrix} \lambda_1 & 0 & . & . & . & 0 \\ 0 & \lambda_2 & & & & . \\ . & & . & & & . \\ . & & & . & & . \\ . & & & & . & . \\ 0 & . & . & . & . & \lambda_n \end{pmatrix} \begin{pmatrix} y_1 \\ y_2 \\ . \\ . \\ . \\ y_n \end{pmatrix} \tag{106}$$

$$= \lambda_1 y_1^2 + \lambda_2 y_2^2 + \ldots + \lambda_n y_n^2. \tag{107}$$

Now as we have seen in 6.1 one of the important problems associated with a quadratic form is to determine the nature of the form – that is, whether it is positive definite, negative definite or indefinite. The results of this section enable this to be decided very easily. For from (107) if all the eigenvalues λ_i of $\mathbf{C}$ are positive then $Q(x_i) > 0$ for all x_i except $x_i = 0$ and consequently is positive definite. Similarly if all the eigenvalues of $\mathbf{C}$ are negative then $Q(x_i) < 0$ for all x_i except $x_i = 0$ and accordingly is negative definite. If, however, $\mathbf{C}$ has both positive and negative eigenvalues then $Q(x_i)$ takes on positive and negative values for different x_i values and is consequently an indefinite form. For example,

$$Q_1 = x_1^2+2x_2^2+3x_3^2 \tag{108}$$

is a positive definite form, and

$$Q_2 = x_1^2+2x_2^2-3x_3^2 \tag{109}$$

is an indefinite form.

An important number associated with a quadratic form is its signature s. This is defined as the number of positive terms minus the number of negative terms in the diagonalised form of Q. By inspection the signatures of Q_1 and Q_2 of (108) and (109) are respectively $+3$ and $+1$, whereas, for example,

$$Q_3 = x_1^2-x_2^2-x_3^2-x_4^2 \tag{110}$$

has signature -2.

An important result is that given two or more real linear transformations which diagonalise a quadratic form the resulting diagonalised forms (although different) nevertheless have the same signature. In other words, the signature is an invariant quantity under real transformations of the variables x_i. (Clearly signature is not an invariant quantity under complex transformations. For example, the transformation $x_1 = y_1$, $x_2 = iy_2$, $x_3 = iy_3$ transforms Q_1 of (108) which has signature $+3$, into $y_1^2-2y_2^2-3y_3^2$, which has signature -1.)

Example 7. The real quadratic form

$$Q = 2x_1^2+2x_2^2+2x_1x_2+3x_3^2 \tag{111}$$

may be written as $\tilde{\mathbf{X}}\mathbf{A}\mathbf{X}$, where

$$\mathbf{A} = \begin{pmatrix} 2 & 1 & 0 \\ 1 & 2 & 0 \\ 0 & 0 & 3 \end{pmatrix} \quad \text{and} \quad \mathbf{X} = \begin{pmatrix} x_1 \\ x_2 \\ x_3 \end{pmatrix}. \tag{112}$$

It is easily found that the eigenvalues of **A** are $\lambda_1 = 1$, $\lambda_2 = \lambda_3 = 3$, and that a corresponding set of orthonormal eigenvectors is

$$\mathbf{q}_1 = \begin{pmatrix} \frac{1}{\sqrt{2}} \\ -\frac{1}{\sqrt{2}} \\ 0 \end{pmatrix}, \quad \mathbf{q}_2 = \begin{pmatrix} \frac{1}{\sqrt{2}} \\ \frac{1}{\sqrt{2}} \\ 0 \end{pmatrix}, \quad \mathbf{q}_3 = \begin{pmatrix} 0 \\ 0 \\ 1 \end{pmatrix}. \tag{113}$$

(*N.B.* $\mathbf{q}_i$ has been used to denote the ith eigenvector of **A** rather than $\mathbf{X}_i$ so as to avoid possible confusion with **X** in (112), which is an *arbitrary* column vector associated with the quadratic form.)

The matrix **A** may now be diagonalised by the orthogonal matrix **U** of the eigenvectors (113), where

$$\mathbf{U} = \begin{pmatrix} \frac{1}{\sqrt{2}} & \frac{1}{\sqrt{2}} & 0 \\ -\frac{1}{\sqrt{2}} & \frac{1}{\sqrt{2}} & 0 \\ 0 & 0 & 1 \end{pmatrix} \tag{114}$$

to give

$$\mathbf{U}^{-1}\mathbf{A}\mathbf{U} = \tilde{\mathbf{U}}\mathbf{A}\mathbf{U} = \mathbf{D} = \begin{pmatrix} 1 & 0 & 0 \\ 0 & 3 & 0 \\ 0 & 0 & 3 \end{pmatrix}. \tag{115}$$

Hence the transformation $\mathbf{X} = \mathbf{U}\mathbf{Y}$, where

$$\mathbf{Y} = \begin{pmatrix} y_1 \\ y_2 \\ y_3 \end{pmatrix}, \tag{116}$$

diagonalises the quadratic form $\tilde{\mathbf{X}}\mathbf{A}\mathbf{X}$ to give

$$\widetilde{(\mathbf{U}\mathbf{Y})}\mathbf{A}\mathbf{U}\mathbf{Y} = \tilde{\mathbf{Y}}\tilde{\mathbf{U}}\mathbf{A}\mathbf{U}\mathbf{Y} = \tilde{\mathbf{Y}}\mathbf{D}\mathbf{Y} = y_1^2+3y_2^2+3y_3^2. \tag{117}$$

Alternatively Q of (111) may be reduced to diagonal form by the Lagrange method of 6.8 to give

$$= 2\left(x_1 + \frac{x_2}{2}\right)^2 + \tfrac{3}{2}x_2^2 + 3x_3^2 \tag{118}$$

$$= 2u_1^2 + \tfrac{3}{2}u_2^2 + 3u_3^2, \tag{119}$$

where

$$\left.\begin{aligned} u_1 &= x_1+\frac{x_2}{2} & \qquad x_1 &= u_1-\frac{u_2}{2}, \\ u_2 &= x_2, \qquad \text{or} & x_2 &= u_2, \\ u_3 &= x_3, & x_3 &= u_3. \end{aligned}\right\} \tag{120}$$

Finally the transformation $\mathbf{X} = \mathbf{WZ}$, where

$$\mathbf{Z} = \begin{pmatrix} z_1 \\ z_2 \\ z_3 \end{pmatrix} \quad \text{and} \quad \mathbf{W} = \tfrac{1}{3}\begin{pmatrix} -1 & 2 & 1 \\ 2 & -1 & 1 \\ 1 & 1 & -1 \end{pmatrix} \tag{121}$$

sends $\tilde{\mathbf{X}}\mathbf{AX}$ into

$$\tilde{\mathbf{Z}}\tilde{\mathbf{W}}\mathbf{AWZ} = z_1^2+z_2^2+z_3^2. \tag{122}$$

Here $\mathbf{W}$ is not an orthogonal matrix – nevertheless $\tilde{\mathbf{W}}\mathbf{AW}$ is diagonal.

In the three diagonalised forms of (111) – namely, (117), (119) and (122) – the coefficients of the variables are different, as are the relations between the x_i and the transformed coordinates. However, the signature of each form is $+3$.

6.10 Hermitian forms

Results similar to those obtained in the last section for a real quadratic form hold for a Hermitian form H defined by

$$H = \widetilde{\mathbf{X}^*}\mathbf{AX} = \mathbf{X}^\dagger\mathbf{AX}, \tag{123}$$

where $\mathbf{A}$ is a Hermitian matrix (i.e. $\mathbf{A}^\dagger = \mathbf{A}$) and $\mathbf{X}$ is a column vector of complex elements. If $\mathbf{A}$ and $\mathbf{X}$ are real then H is a real quadratic form. One of the important properties of a Hermitian form is that its value is always real. This is easily proved by considering

$$H^* = (\widetilde{\mathbf{X}^*}\mathbf{AX})^* = \tilde{\mathbf{X}}\mathbf{A}^*\mathbf{X}^* = \tilde{\mathbf{X}}\tilde{\mathbf{A}}\mathbf{X}^*. \tag{124}$$

Now, since the transpose of a number is itself,

$$(\tilde{\mathbf{X}}\tilde{\mathbf{A}}\mathbf{X}^*) = (\widetilde{\tilde{\mathbf{X}}\tilde{\mathbf{A}}\mathbf{X}^*}) = \widetilde{\mathbf{X}^*}\mathbf{AX} = H. \tag{125}$$

Consequently

$$H^* = H, \tag{126}$$

showing that H is real.

A Hermitian form may be diagonalised in a similar way to a real quadratic form. Consider the non-singular complex linear transformation

$$\mathbf{X} = \mathbf{UY}, \tag{127}$$

applied to H so that H becomes

$$(\mathbf{UY})^\dagger\mathbf{AUY} = \mathbf{Y}^\dagger\mathbf{U}^\dagger\mathbf{AUY}. \tag{128}$$

If now $\mathbf{U}$ is chosen to be the unitary matrix formed from the column eigenvectors of $\mathbf{A}$ (see 6.6) then

$$\mathbf{U}^\dagger\mathbf{AU}(=\widetilde{\mathbf{U}^*}\mathbf{AU}) = \mathbf{D}, \tag{129}$$

where $\mathbf{D}$ is a diagonal matrix with the eigenvalues (real) of $\mathbf{A}$ as elements. Consequently by (128) and (129) we have

$$H = \mathbf{Y}^\dagger\mathbf{DY} = \lambda_1|y_1|^2+\lambda_2|y_2|^2+\ldots+\lambda_n|y_n|^2, \tag{130}$$

where $|y_1|$ denotes the modulus $\sqrt{(y_1^*y_1)}$ of y_1, etc., and $\lambda_1, \lambda_2, \ldots, \lambda_n$ are the eigenvalues of $\mathbf{A}$. (Since the eigenvalues of a Hermitian matrix are necessarily real, (130) again demonstrates that H is a real quantity.)

6.11 Simultaneous diagonalisation of two quadratic forms

It has been shown in 6.5 that a real symmetric matrix may always be diagonalised by an orthogonal transformation. We now wish to find the conditions under which two real symmetric matrices may be diagonalised by the same orthogonal transformation. Suppose the real symmetric matrix $\mathbf{A}$ is diagonalised by the orthogonal matrix $\mathbf{U}(\tilde{\mathbf{U}} = \mathbf{U}^{-1})$ so that

$$\mathbf{U}^{-1}\mathbf{AU} = \tilde{\mathbf{U}}\mathbf{AU} = \mathbf{D}_1. \tag{131}$$

Now if $\mathbf{B}$ is another real symmetric matrix, then in general $\tilde{\mathbf{U}}\mathbf{BU}$ will not be a diagonal matrix. However, *if* $\tilde{\mathbf{U}}\mathbf{BU}$ is a diagonal matrix $\mathbf{D}_2$ (say) then, since $\mathbf{D}_1\mathbf{D}_2 = \mathbf{D}_2\mathbf{D}_1$, we have

$$\tilde{\mathbf{U}}\mathbf{AU}\tilde{\mathbf{U}}\mathbf{BU} = \tilde{\mathbf{U}}\mathbf{BU}\tilde{\mathbf{U}}\mathbf{AU}. \tag{132}$$

Now, since $\mathbf{U}$ is orthogonal, $\mathbf{U}\tilde{\mathbf{U}} = \mathbf{I}$, and hence

$$\mathbf{AB} = \mathbf{BA}. \tag{133}$$

In other words, if two real symmetric matrices are simultaneously diagonalisable by the same orthogonal transformation they must, of necessity, commute. It may also be shown that this condition is sufficient in that if two real symmetric matrices commute then they may be simultaneously diagonalised by the same *orthogonal* transformation.

It follows that two real quadratic forms $Q_1 = \tilde{\mathbf{X}}\mathbf{AX}$, $Q_2 = \tilde{\mathbf{X}}\mathbf{BX}$ A and B symmetric) may be simultaneously reduced to diagonal form by the same *orthogonal* transformation only if **A** and **B** commute.

In quantum mechanics special importance is attached to unitary transformations. It is easily seen that if **H** is now a Hermitian matrix which is diagonalised by a unitary matrix $\mathbf{U}(\mathbf{U}^\dagger = \widetilde{\mathbf{U}^*} = \mathbf{U}^{-1})$ so that $\mathbf{U}^\dagger\mathbf{H}\mathbf{U}$ is diagonal, then another Hermitian matrix **K** is diagonalised by the same unitary transformation only if **H** and **K** commute. In the same way as for real quadratic forms it follows that two Hermitian forms $H_1 = \mathbf{X}^\dagger\mathbf{H}\mathbf{X}$, $H_2 = \mathbf{X}^\dagger\mathbf{K}\mathbf{X}$ may be simultaneously reduced to diagonal form by the same unitary transformation only if **H** and **K** commute.

Example 8. The symmetric matrices

$$\mathbf{A} = \begin{pmatrix} 2 & 1 \\ 1 & 2 \end{pmatrix}, \qquad \mathbf{B} = \begin{pmatrix} 3 & 2 \\ 2 & 3 \end{pmatrix} \tag{134}$$

commute. Hence they can both be diagonalised by the same orthogonal transformation. Now the orthogonal matrix formed from the orthonormal eigenvectors of **A** is easily found to be

$$\mathbf{U} = \begin{pmatrix} \frac{1}{\sqrt{2}} & \frac{1}{\sqrt{2}} \\ \frac{1}{\sqrt{2}} & -\frac{1}{\sqrt{2}} \end{pmatrix}. \tag{135}$$

Consequently

$$\tilde{\mathbf{U}}\mathbf{A}\mathbf{U} = \begin{pmatrix} 3 & 0 \\ 0 & 1 \end{pmatrix} \quad \text{and} \quad \tilde{\mathbf{U}}\mathbf{B}\mathbf{U} = \begin{pmatrix} 5 & 0 \\ 0 & 1 \end{pmatrix}, \tag{136}$$

showing that both **A** and **B** are diagonalised by **U**.

PROBLEMS 6

1. Diagonalise each of the following matrices by means of a similarity transformation:

   (a) $\begin{pmatrix} 2 & 3 \\ 4 & 1 \end{pmatrix}$, (b) $\begin{pmatrix} 1 & 0 & -1 \\ 1 & 2 & 1 \\ 2 & 2 & 3 \end{pmatrix}$, (c) $\begin{pmatrix} -2 & -1 & 0 \\ 1 & 2 & 3 \\ 4 & 5 & 6 \end{pmatrix}$.

   In each case obtain the sixth power of the matrix by first finding the sixth power of the diagonal form and then transforming back.

2. Diagonalise each of the following real symmetric matrices by means of an orthogonal transformation:

(a) $\begin{pmatrix} 3 & 4 \\ 4 & -3 \end{pmatrix}$, (b) $\begin{pmatrix} 0 & 1 & 0 \\ 1 & 0 & 0 \\ 0 & 0 & 1 \end{pmatrix}$,

(c) $\begin{pmatrix} 2 & 2 & 0 \\ 2 & 2 & 0 \\ 0 & 0 & 1 \end{pmatrix}$, (d) $\begin{pmatrix} 0 & 1 & 0 & 0 \\ 1 & 0 & 0 & 0 \\ 0 & 0 & 0 & 1 \\ 0 & 0 & 1 & 0 \end{pmatrix}$.

3. Diagonalise each of the following Hermitian matrices by means of a unitary transformation:

(a) $\begin{pmatrix} -2 & 3+3i \\ 3-3i & 1 \end{pmatrix}$, (b) $\begin{pmatrix} 3 & 1-i \\ 1+i & 4 \end{pmatrix}$,

(c) $\begin{pmatrix} 0 & i & 0 \\ -i & 4 & -2i \\ 0 & 2i & 2 \end{pmatrix}$.

4. Find the real symmetric matrix associated with each of the following quadratic forms:

(a) $2x_1^2-5x_1x_2+5x_2^2$,

(b) $x_1^2-2x_1x_2+2x_2^2-2x_2x_3+2x_3^2$,

(c) $x_1^2+8x_1x_2-10x_1x_3+2x_3^2$.

5. Reduce each of the above quadratic forms to diagonal form by Lagrange's method.

6. It can be proved that a set of necessary and sufficient conditions for the quadratic form $\tilde{\mathbf{X}}\mathbf{A}\mathbf{X}$ to be positive definite, where $\mathbf{A}$ is a real symmetric matrix with elements a_{ik}, is that all the determinants

$$A_1 = a_{11}, \quad A_2 = \begin{vmatrix} a_{11} & a_{12} \\ a_{21} & a_{22} \end{vmatrix},$$

$$A_3 = \begin{vmatrix} a_{11} & a_{12} & a_{13} \\ a_{21} & a_{22} & a_{23} \\ a_{31} & a_{32} & a_{33} \end{vmatrix}, \ \ldots, \quad A_n = |\mathbf{A}|$$

are positive.

Using these conditions, show that the quadratic forms

$$3x_1^2+4x_1x_2+5x_2^2$$

and

$$2x_1^2+2x_1x_2+2x_2^2-6x_1x_3-2x_2x_3+7x_3^2$$

are both positive definite.

7. Suppose that the real quadratic form $\tilde{\mathbf{X}}\mathbf{AX}$ is diagonalised by the orthogonal transformation $\mathbf{X} = \mathbf{UY}$ ($\mathbf{U}$ orthogonal) to give

$$\tilde{\mathbf{Y}}\mathbf{DY} = \lambda_1 y_1^2+\lambda_2 y_2^2+\ldots+\lambda_n y_n^2,$$

where $\lambda_1, \lambda_2, \ldots, \lambda_n$ are the eigenvalues of $\mathbf{A}$. We now assume that $\lambda_1 \geq \lambda_2 \geq \ldots \geq \lambda_n$. Then

$$\tilde{\mathbf{Y}}\mathbf{DY} \geq \lambda_n \tilde{\mathbf{Y}}\mathbf{Y} \quad \text{and} \quad \tilde{\mathbf{Y}}\mathbf{DY} \leq \lambda_1 \tilde{\mathbf{Y}}\mathbf{Y}.$$

Hence

$$\lambda_1 = \max.\frac{\tilde{\mathbf{X}}\mathbf{AX}}{\tilde{\mathbf{X}}\mathbf{X}}, \qquad \lambda_n = \min.\frac{\tilde{\mathbf{X}}\mathbf{AX}}{\tilde{\mathbf{X}}\mathbf{X}},$$

and consequently

$$\lambda_1 \geq \frac{\tilde{\mathbf{X}}\mathbf{AX}}{\tilde{\mathbf{X}}\mathbf{X}} \geq \lambda_n.$$

By choosing different forms for the vector $\mathbf{X}$, obtain approximate bounds on the eigenvalues of

$$\mathbf{A} = \begin{pmatrix} 2 & 1 & 0 \\ 1 & 3 & 0 \\ 0 & 0 & 2 \end{pmatrix}.$$

CHAPTER 7

Functions of Matrices

7.1 Introduction

In the last chapter it was shown that powers of a square matrix could readily be obtained by putting the matrix into diagonal form by means of a similarity transformation. Powers of matrices are frequently required especially in the study of matrix functions, where for example a power series such as

$$e^{\mathbf{A}} = \mathbf{I} + \mathbf{A} + \frac{\mathbf{A}^2}{2\,!} + \dots \tag{1}$$

depends on all positive powers of **A**. We shall deal with matrix functions – in particular, with (1) – in later sections of this chapter. For the moment, however, we recall that not all matrices are diagonalisable by a similarity transformation and consequently some other method is required for evaluating powers of matrices. Such a method, which in fact can be applied to all (square) matrices, is embodied in the Cayley-Hamilton theorem discussed in the next section.

7.2 Cayley-Hamilton theorem

This theorem states that every square matrix satisfies its own characteristic equation. In other words if

$$f(\lambda) = |\mathbf{A} - \lambda \mathbf{I}| \tag{2}$$

is the characteristic polynomial of an nth order matrix **A** then

$$f(\mathbf{A}) = \mathbf{0}, \tag{3}$$

where **0** is the zero matrix of order n.

We may see the origin of this theorem in the following analysis. For, by Chapter 5, 5.3,

$$f(\lambda) = (-1)^n(\lambda^n - \alpha_1 \lambda^{n-1} + \alpha_2 \lambda^{n-2} - \dots + (-1)^n \alpha_n) \tag{4}$$

and hence

$$f(\mathbf{A}) = (-1)^n(\mathbf{A}^n - \alpha_1 \mathbf{A}^{n-1} + \alpha_2 \mathbf{A}^{n-2} - \dots + (-1)^n \alpha_n \mathbf{I}). \tag{5}$$

Now if $\mathbf{X}_i$ is an eigenvector of $\mathbf{A}$ corresponding to the ith eigenvalue λ_i then

$$\mathbf{AX}_i = \lambda_i \mathbf{X}_i \tag{6}$$

and

$$f(\mathbf{A})\mathbf{X}_i = (-1)^n(\mathbf{A}^n - \alpha_1 \mathbf{A}^{n-1} + \alpha_2 \mathbf{A}^{n-2} \ldots + (-1)^n \alpha_n \mathbf{I})\mathbf{X}_i. \tag{7}$$

However, from (6) it follows that

$$\mathbf{A}^n \mathbf{X}_i = \lambda_i^n \mathbf{X}_i. \tag{8}$$

Hence using (8) in each term on the right-hand side of (7) we have

$$f(\mathbf{A})\mathbf{X}_i = (-1)^n(\lambda_i^n - \alpha_1 \lambda_i^{n-1} + \alpha_2 \lambda_i^{n-2} \ldots + (-1)^n \alpha_n)\mathbf{X}_i \tag{9}$$

$$= \mathbf{0}, \text{ since } \lambda_i \text{ is an eigenvalue of } \mathbf{A}. \tag{10}$$

Now if $\mathbf{A}$ has n distinct (i.e. not repeated) eigenvalues there will be n linearly independent eigenvectors $\mathbf{X}_1, \mathbf{X}_2, \ldots, \mathbf{X}_n$. Writing the matrix of these eigenvectors as

$$\mathbf{S} = (\mathbf{X}_1 \mathbf{X}_2 \ldots \mathbf{X}_n), \tag{11}$$

(10) may be expressed as

$$f(\mathbf{A})\mathbf{S} = \mathbf{0}. \tag{12}$$

Since the $\mathbf{X}_i$ are linearly independent, $\mathbf{S}^{-1}$ exists, and hence

$$f(\mathbf{A})\mathbf{SS}^{-1} = \mathbf{0} \tag{13}$$

or

$$f(\mathbf{A}) = \mathbf{0}, \tag{14}$$

which is the Cayley-Hamilton theorem.

This proof depends on $\mathbf{A}$ having distinct eigenvalues. We now show that the Cayley-Hamilton theorem is true for any n^{th} order $\mathbf{A}$ whether it has repeated eigenvalues or not.

Consider first $\operatorname{adj}(\mathbf{A} - \lambda\mathbf{I})$ which is, by definition, the transposed matrix of the cofactors of $\mathbf{A} - \lambda\mathbf{I}$. Now, since $|\mathbf{A} - \lambda\mathbf{I}|$ is a polynomial of degree n in λ, $\operatorname{adj}(\mathbf{A} - \lambda\mathbf{I})$ will, in general, be a polynomial of degree $n-1$ in λ with matrix coefficients. Hence we may write

$$\operatorname{adj}(\mathbf{A} - \lambda\mathbf{I}) = \mathbf{C}_0 \lambda^{n-1} + \mathbf{C}_1 \lambda^{n-2} + \ldots + \mathbf{C}_{n-1}, \tag{15}$$

where $\mathbf{C}_0, \mathbf{C}_1, \ldots, \mathbf{C}_{n-1}$ are n^{th} order matrices with elements dependent on the elements of $\mathbf{A}$.

Now by definition

$$(\mathbf{A} - \lambda\mathbf{I}) \operatorname{adj}(\mathbf{A} - \lambda\mathbf{I}) = |\mathbf{A} - \lambda\mathbf{I}|\mathbf{I} = f(\lambda)\mathbf{I}. \tag{16}$$

Hence

$$(\mathbf{A} - \lambda\mathbf{I})(\mathbf{C}_0 \lambda^{n-1} + \mathbf{C}_1 \lambda^{n-2} + \ldots + \mathbf{C}_{n-1}) = (-1)^n \times$$
$$(\lambda^n - \alpha_1 \lambda^{n-1} + \alpha_2 \lambda^{n-2} \ldots + (-1)^n \alpha_n)\mathbf{I}. \tag{17}$$

Equating coefficients of like powers of λ on each side of (17) we find

$$\left.\begin{aligned} -\mathbf{C}_0 &= (-1)^n\mathbf{I}, \\ \mathbf{AC}_0-\mathbf{C}_1 &= -(-1)^n\alpha_1\mathbf{I}, \\ \mathbf{AC}_1-\mathbf{C}_2 &= (-1)^n\alpha_2\mathbf{I}, \\ &\;\;\vdots \\ \mathbf{AC}_{n-1} &= \alpha_n\mathbf{I}. \end{aligned}\right\} \tag{18}$$

Pre-multiplying the first equation in (18) by $\mathbf{A}^n$, the second by $\mathbf{A}^{n-1}$, and so on, the last being pre-multiplied by $\mathbf{I}$, and then finally adding the resulting equations we have

$$\mathbf{0} = (-1)^n(\mathbf{A}^n-\alpha_1\mathbf{A}^{n-1}+\alpha_2\mathbf{A}^{n-2}+\ldots+(-1)^n\alpha_n\mathbf{I}) \tag{19}$$

or

$$f(\mathbf{A}) = \mathbf{0}, \tag{20}$$

which again is the Cayley-Hamilton theorem.

The following examples illustrate the theorem.

Example 1. Consider the matrix

$$\mathbf{A} = \begin{pmatrix} 1 & 2 \\ 4 & 3 \end{pmatrix} \tag{21}$$

whose characteristic equation ($|\mathbf{A}-\lambda\mathbf{I}| = 0$) is

$$\lambda^2-4\lambda-5 = 0. \tag{22}$$

Hence, by the Cayley-Hamilton theorem, $\mathbf{A}$ must satisfy the relation

$$\mathbf{A}^2-4\mathbf{A}-5\mathbf{I} = \mathbf{0}, \tag{23}$$

where $\mathbf{0}$ is the zero matrix of order 2. This is easily verified by evaluating $\mathbf{A}^2$ directly to give

$$\mathbf{A}^2 = \begin{pmatrix} 9 & 8 \\ 16 & 17 \end{pmatrix} = 4\begin{pmatrix} 1 & 2 \\ 4 & 3 \end{pmatrix} + 5\begin{pmatrix} 1 & 0 \\ 0 & 1 \end{pmatrix}. \tag{24}$$

To evaluate $\mathbf{A}^3$ we write (using (23))

$$\begin{aligned} \mathbf{A}^3 = \mathbf{AA}^2 &= \mathbf{A}(4\mathbf{A}+5\mathbf{I}) = 4\mathbf{A}^2+5\mathbf{A} \\ &= 5\mathbf{A}+4(4\mathbf{A}+5\mathbf{I}) = 21\mathbf{A}+20\mathbf{I} \qquad (25) \\ &= 21\begin{pmatrix} 1 & 2 \\ 4 & 3 \end{pmatrix} + 20\begin{pmatrix} 1 & 0 \\ 0 & 1 \end{pmatrix} = \begin{pmatrix} 41 & 42 \\ 84 & 83 \end{pmatrix}. \qquad (26) \end{aligned}$$

Likewise

$$\mathbf{A}^4 = \mathbf{A}\mathbf{A}^3 = \mathbf{A}(21\mathbf{A}+20\mathbf{I}) \quad \text{(using (25))} \tag{27}$$

$$= 21\mathbf{A}^2+20\mathbf{A} = 21(4\mathbf{A}+5\mathbf{I})+20\mathbf{A} \quad \text{(using (23))} \tag{28}$$

$$= 104\mathbf{A}+105\mathbf{I} \tag{29}$$

$$= 104\begin{pmatrix} 1 & 2 \\ 4 & 3 \end{pmatrix} + 105\begin{pmatrix} 1 & 0 \\ 0 & 1 \end{pmatrix} = \begin{pmatrix} 209 & 208 \\ 416 & 417 \end{pmatrix}. \tag{30}$$

Furthermore, since $\mathbf{A}$ is non-singular, (23) may be written as

$$\mathbf{A}-4\mathbf{I}-5\mathbf{A}^{-1} = \mathbf{0} \tag{31}$$

so that

$$\mathbf{A}^{-1} = \tfrac{1}{5}(\mathbf{A}-4\mathbf{I}) = \tfrac{1}{5}\begin{pmatrix} 1 & 2 \\ 4 & 3 \end{pmatrix} - \tfrac{4}{5}\begin{pmatrix} 1 & 0 \\ 0 & 1 \end{pmatrix} \tag{32}$$

$$= \begin{pmatrix} -\tfrac{3}{5} & \tfrac{2}{5} \\ \tfrac{4}{5} & -\tfrac{1}{5} \end{pmatrix}. \tag{33}$$

This is a very useful way of evaluating the inverse of a matrix, and may be readily extended to higher negative powers. For example,

$$\mathbf{A}^{-2} = \mathbf{A}^{-1}\mathbf{A}^{-1} = \tfrac{1}{5}(\mathbf{A}-4\mathbf{I})\tfrac{1}{5}(\mathbf{A}-4\mathbf{I})$$

$$= \tfrac{1}{25}(\mathbf{A}^2-8\mathbf{A}+16\mathbf{I})$$

$$= \tfrac{1}{25}(4\mathbf{A}+5\mathbf{I})-\tfrac{8}{25}\mathbf{A}+\tfrac{16}{25}\mathbf{I}, \quad \text{(using (23))},$$

$$= -\tfrac{4}{25}\mathbf{A}+\tfrac{21}{25}\mathbf{I}, \tag{34}$$

$$= -\tfrac{4}{25}\begin{pmatrix} 1 & 2 \\ 4 & 3 \end{pmatrix} + \tfrac{21}{25}\begin{pmatrix} 1 & 0 \\ 0 & 1 \end{pmatrix}$$

$$= \begin{pmatrix} \tfrac{17}{25} & -\tfrac{8}{25} \\ -\tfrac{16}{25} & \tfrac{9}{25} \end{pmatrix}, \tag{35}$$

as may be verified from first principles.

Clearly all positive and negative integral powers of $\mathbf{A}$ may be expressed as linear combinations of $\mathbf{A}$ itself and the unit matrix $\mathbf{I}$; that is

$$\mathbf{A}^r = a_1\mathbf{A}+a_2\mathbf{I}, \tag{36}$$

where r is a positive or negative integer, and a_1 and a_2 are numerical constants which are different for each r (as shown by (25), (29), (32) and (34)). This result is true for any matrix of order 2 (assuming that it is non-singular for the negative powers to exist) and in 7.3 we show how the constants a_1 and a_2 may be evaluated.

Example 2. The matrix

$$\mathbf{A} = \begin{pmatrix} 2 & 1 & 2 \\ 0 & 2 & 3 \\ 0 & 0 & 5 \end{pmatrix} \tag{37}$$

has as characteristic equation

$$\lambda^3 - 9\lambda^2 + 24\lambda - 20 = 0. \tag{38}$$

Hence, by the Cayley-Hamilton theorem

$$\mathbf{A}^3 - 9\mathbf{A}^2 + 24\mathbf{A} - 20\mathbf{I} = \mathbf{0}. \tag{39}$$

Consequently, $\mathbf{A}^4$, for example, may be evaluated by writing

$$\mathbf{A}^4 = \mathbf{A}\mathbf{A}^3 = \mathbf{A}(9\mathbf{A}^2 - 24\mathbf{A} + 20\mathbf{I}) \tag{40}$$

$$= 9(9\mathbf{A}^2 - 24\mathbf{A} + 20\mathbf{I}) - 24\mathbf{A}^2 + 20\mathbf{A} \tag{41}$$

$$= 57\mathbf{A}^2 - 196\mathbf{A} + 180\mathbf{I} \tag{42}$$

$$= 57\begin{pmatrix} 4 & 4 & 17 \\ 0 & 4 & 21 \\ 0 & 0 & 25 \end{pmatrix} - 196\begin{pmatrix} 2 & 1 & 2 \\ 0 & 2 & 3 \\ 0 & 0 & 5 \end{pmatrix} + 180\begin{pmatrix} 1 & 0 & 0 \\ 0 & 1 & 0 \\ 0 & 0 & 1 \end{pmatrix}$$

$$= \begin{pmatrix} 16 & 32 & 577 \\ 180 & 16 & 609 \\ 0 & 0 & 625 \end{pmatrix}. \tag{43}$$

Moreover, since $\mathbf{A}$ is non-singular, $\mathbf{A}^{-1}$ may be evaluated by writing (39) as

$$\mathbf{A}^2 - 9\mathbf{A} + 24\mathbf{I} - 20\mathbf{A}^{-1} = \mathbf{0} \tag{44}$$

or

$$\mathbf{A}^{-1} = \tfrac{1}{20}(\mathbf{A}^2 - 9\mathbf{A} + 24\mathbf{I}), \tag{45}$$

$$= \tfrac{1}{20}\begin{pmatrix} 10 & -5 & -1 \\ 0 & 10 & -6 \\ 0 & 0 & 4 \end{pmatrix}. \tag{46}$$

Similar calculations can be made for higher positive and negative integral powers of $\mathbf{A}$. In fact, it is clear (see, for example, (42) and (45)) that

$$\mathbf{A}^r = a_1\mathbf{A}^2 + a_2\mathbf{A} + a_3\mathbf{I}, \tag{47}$$

where r is a positive or negative integer, and a_1, a_2 and a_3 are numerical constants whose values depend on the value of r.

7.3 Powers of matrices

The previous two examples have shown that (a) in the case of a second order (i.e. (2×2)) matrix $\mathbf{A}$,

$$\mathbf{A}^r = a_1\mathbf{A}+a_2\mathbf{I}, \tag{48}$$

and (b) in the case of a third order matrix $\mathbf{A}$,

$$\mathbf{A}^r = a_1\mathbf{A}^2+a_2\mathbf{A}+a_3\mathbf{I}. \tag{49}$$

Using the Cayley-Hamilton theorem it is easily found that for an nth order matrix any integral power $\mathbf{A}^r$ may be expressed as

$$\mathbf{A}^r = a_1\mathbf{A}^{n-1}+a_2\mathbf{A}^{n-2}+a_3\mathbf{A}^{n-3}+\ldots+a_{n-1}\mathbf{A}+a_n\mathbf{I}, \tag{50}$$

where the values of $a_1, a_2, \ldots, a_n$ depend on the particular choice of r. However, as Example 2 showed, the evaluation of the constants $a_1, a_2, \ldots$ etc., required repeated use of the Cayley-Hamilton theorem. We give here an alternative procedure for the calculation of these constants. For convenience we deal only with second order matrices, the analysis for n^{th} order matrices (which is exactly similar) being partly discussed in Problems 4 and 5 at the end of the chapter.

Now the characteristic polynomial $f(\lambda)$ of a second-order matrix $\mathbf{A}$ is a quadratic expression. Hence if λ^r is divided by $f(\lambda)$ we have

$$\lambda^r = f(\lambda)Q(\lambda)+R(\lambda), \tag{51}$$

where $Q(\lambda)$ is a quotient polynomial and $R(\lambda)$ is a remainder polynomial which at most is of first degree.

Let

$$R(\lambda) = a_1\lambda+a_2. \tag{52}$$

Then since the eigenvalues of $\mathbf{A}$, λ_1 and λ_2, say, are the roots of $f(\lambda) = 0$, we have, using (51) and (52)

$$\left.\begin{aligned} \lambda_1^r &= a_1\lambda_1+a_2,\\ \lambda_2^r &= a_1\lambda_2+a_2.\end{aligned}\right\} \tag{53}$$

Provided $\lambda_1 \neq \lambda_2$, the two equations of (53) determine the values of a_1 and a_2.

Now the analogous result to (51) for the matrix $\mathbf{A}$ (which we state here without proof) is

$$\mathbf{A}^r = f(\mathbf{A})Q(\mathbf{A})+R(\mathbf{A}). \tag{54}$$

However, by the Cayley-Hamilton theorem $f(\mathbf{A}) = \mathbf{0}$. Hence

$$\mathbf{A}^r = R(\mathbf{A}) = a_1\mathbf{A}+a_2\mathbf{I}, \tag{55}$$

which is precisely the result of (48). The values of a_1 and a_2, however, are now the solutions of (53).

Example 3. We consider the matrix

$$\mathbf{A} = \begin{pmatrix} 1 & 2 \\ 4 & 3 \end{pmatrix} \tag{56}$$

of Example 1. To evaluate $\mathbf{A}^3$ we write

$$\mathbf{A}^3 = a_1 \mathbf{A} + a_2 \mathbf{I}, \tag{57}$$

and determine a_1 and a_2 from the equations (53) making use of the fact that the eigenvalues of $\mathbf{A}$ are $\lambda_1 = 5$ and $\lambda_2 = -1$. That is

$$\left.\begin{aligned} 5^3 &= 5a_1 + a_2, \\ (-1)^3 &= -a_1 + a_2, \end{aligned}\right\} \tag{58}$$

whence

$$a_1 = 21, \qquad a_2 = 20. \tag{59}$$

Hence

$$\mathbf{A}^3 = 21\mathbf{A} + 20\mathbf{I} \tag{60}$$

$$= 21 \begin{pmatrix} 1 & 2 \\ 4 & 3 \end{pmatrix} + 20 \begin{pmatrix} 1 & 0 \\ 0 & 1 \end{pmatrix} = \begin{pmatrix} 41 & 42 \\ 84 & 83 \end{pmatrix} \tag{61}$$

as in (26).

Likewise

$$\mathbf{A}^6 = b_1 \mathbf{A} + b_2 \mathbf{I}, \quad \text{(say)}, \tag{62}$$

where b_1 and b_2 are the solutions of

$$\left.\begin{aligned} 5^6 &= 5b_1 + b_2, \\ (-1)^6 &= -b_1 + b_2. \end{aligned}\right\} \tag{63}$$

Solving (63) we find

$$b_1 = 2771, \qquad b_2 = 2770. \tag{64}$$

Hence

$$\mathbf{A}^6 = 2771 \begin{pmatrix} 1 & 2 \\ 3 & 4 \end{pmatrix} + 2770 \begin{pmatrix} 1 & 0 \\ 0 & 1 \end{pmatrix} = \begin{pmatrix} 5541 & 5542 \\ 8313 & 13854 \end{pmatrix}. \tag{65}$$

The eigenvalues of $\mathbf{A}$ in the last example are different, this being the condition under which (53) leads to unique values of a_1 and a_2. It is natural to ask how these constants can be determined for a matrix (again second order) with two identical eigenvalues. Suppose λ_1 is a double root of the characteristic equation $f(\lambda) = 0$. Then

$$f(\lambda_1) = 0, \qquad \text{and } f'(\lambda_1) = 0. \tag{66}$$

Now differentiating (51) with respect to λ we have

$$r\lambda^{r-1} = f(\lambda)Q'(\lambda) + f'(\lambda)Q(\lambda) + R'(\lambda). \tag{67}$$

Hence putting $\lambda = \lambda_1$ and using (66) we find

$$r\lambda_1^{r-1} = R'(\lambda_1) = a_1. \tag{68}$$

The pair of equations

$$\left.\begin{aligned} \lambda_1^r &= a_1\lambda_1 + a_2, \\ r\lambda_1^{r-1} &= a_1, \end{aligned}\right\} \tag{69}$$

then determine uniquely the values of a_1 and a_2 and take the place of (53) when the two eigenvalues are the same.

Example 4. Consider the matrix

$$\mathbf{A} = \begin{pmatrix} 1 & 2 \\ 0 & 1 \end{pmatrix} \tag{70}$$

which has eigenvalues $\lambda_1 = 1$ (twice).

Hence

$$\mathbf{A}^r = a_1\mathbf{A} + a_2\mathbf{I}, \tag{71}$$

where a_1 and a_2 are the solutions of the equations

$$\left.\begin{aligned} 1^r &= 1a_1 + a_2, \\ r1^{r-1} &= a_1. \end{aligned}\right\} \tag{72}$$

Equation (72) leads directly to

$$a_1 = r, \qquad a_2 = 1 - r. \tag{73}$$

Hence, for example,

$$\mathbf{A}^3 = 3\begin{pmatrix} 1 & 2 \\ 0 & 1 \end{pmatrix} - 2\begin{pmatrix} 1 & 0 \\ 0 & 1 \end{pmatrix} = \begin{pmatrix} 1 & 6 \\ 0 & 1 \end{pmatrix}, \tag{74}$$

$$\mathbf{A}^{26} = 26\begin{pmatrix} 1 & 2 \\ 0 & 1 \end{pmatrix} - 25\begin{pmatrix} 1 & 0 \\ 0 & 1 \end{pmatrix} = \begin{pmatrix} 1 & 52 \\ 0 & 1 \end{pmatrix}, \tag{75}$$

and

$$\mathbf{A}^{-1} = -\begin{pmatrix} 1 & 2 \\ 0 & 1 \end{pmatrix} + 2\begin{pmatrix} 1 & 0 \\ 0 & 1 \end{pmatrix} = \begin{pmatrix} 1 & -2 \\ 0 & 1 \end{pmatrix}. \tag{76}$$

As mentioned earlier the analysis for n^{th} order matrices is similar to that developed here for second order matrices, and further details can be found in Problems 4 and 5 at the end of the chapter.

7.4 Some matrix series

We indicated in 7.1 that matrix functions – in particular, power series – would be considered later. This is an extensive subject and all that is possible here is to give a brief introduction to it without any proofs.

First we recall some ideas relating to series whose arguments are scalar quantities. Suppose z is a complex number. The series

$$\sum_{r=0}^{\infty} a_r z^r,$$

where the a_r are real coefficients, will converge absolutely to a sum $f(z)$ (say) if

$$\lim_{r\to\infty} \left| \frac{a_{r+1} z^{r+1}}{a_r z^r} \right| < 1 \quad \text{(D'Alembert's ratio test);} \tag{77}$$

that is, if

$$|z| < R, \tag{78}$$

where

$$R = \lim_{r\to\infty} \left| \frac{a_r}{a_{r+1}} \right|. \tag{79}$$

Since z is a complex number, (78) defines a circle of radius R in the Argand plane with centre at the origin (see Fig. 7.1). The series converges for all values of z inside the circle and diverges for all z outside; for this reason, the circle is called the circle of convergence.

For example, the series

$$1 + z + \frac{z^2}{2!} + \ldots + \frac{z^r}{r!} + \ldots = \sum_{r=0}^{\infty} \frac{z^r}{r!} \tag{80}$$

has a circle of convergence of radius

$$R = \lim_{r\to\infty} \left| \frac{1/r!}{1/(r+1)!} \right| = \lim_{r\to\infty} |r+1| = \infty, \tag{81}$$

and consequently converges absolutely for all z. It is easily seen that (80) is just the power series expansion of e^z.

On the other hand the series

$$1+z+z^2+\ldots+z^r+ = \sum_{r=0}^{\infty} z^r \tag{82}$$

has

$$R = \lim_{r\to\infty} \left| \frac{a_r}{a_{r+1}} \right| = \lim_{r\to\infty} \left| \frac{1}{1} \right| = 1, \tag{83}$$

and consequently is absolutely convergent only for $|z| < 1$. For such values of z the series represents the function $\dfrac{1}{1-z}$.

We are now in a position to be able to state (without proof) one of the basic theorems of matrix analysis. This is that if all the eigenvalues of a matrix **A** lie within the circle of convergence of the power series

$$f(z) = \sum_{r=0}^{\infty} a_r z^r \tag{84}$$

then the matrix power series

$$\sum_{r=0}^{\infty} a_r \mathbf{A}^r \tag{85}$$

(where $\mathbf{A}^0$ is defined as the unit matrix **I**) converges absolutely to the matrix function $f(\mathbf{A})$. If at least one eigenvalue of **A** lies outside the circle of convergence, (85) diverges. (A more refined test of convergence is necessary when one or more of the eigenvalues of **A** lies on the circle of convergence – this case will not be discussed here.) For example, since the functions e^z, $\sin z$, $\cos z$ converge for all z (i.e. $R = \infty$), it follows that the matrix functions

$$e^{\mathbf{A}} = \mathbf{I} + \mathbf{A} + \frac{\mathbf{A}^2}{2!} + \ldots + \frac{\mathbf{A}^r}{r!} + \ldots \tag{86}$$

$$\sin \mathbf{A} = \mathbf{A} - \frac{\mathbf{A}^3}{3!} + \frac{\mathbf{A}^5}{5!} - \ldots + (-1)^r \frac{\mathbf{A}^{2r+1}}{(2r+1)!} + \ldots \tag{87}$$

$$\cos \mathbf{A} = \mathbf{I} - \frac{\mathbf{A}^2}{2!} + \frac{\mathbf{A}^4}{4!} - \ldots + (-1)^r \frac{\mathbf{A}^{2r}}{(2r)!} + \ldots \tag{88}$$

are valid for every square matrix **A**.

Likewise

$$e^{i\mathbf{A}} = \cos \mathbf{A} + i \sin \mathbf{A}. \tag{89}$$

Now, since $e^z e^{-z} = 1$, we have

$$e^{\mathbf{A}} e^{-\mathbf{A}} = \mathbf{I} \tag{90}$$

whence

$$(e^{\mathbf{A}})^{-1} = e^{-\mathbf{A}}. \tag{91}$$

Hence the inverse of $e^{\mathbf{A}}$ always exists and $e^{\mathbf{A}}$ is consequently a non-singular matrix for every **A**.

Care must be taken when dealing with more than one matrix function. For, although $e^x e^y = e^{x+y}$, it is not necessarily true that $e^{\mathbf{A}} e^{\mathbf{B}} = e^{\mathbf{A}+\mathbf{B}}$. This may be seen to be the case since (by (86))

$$e^{\mathbf{A}+\mathbf{B}} = \mathbf{I} + (\mathbf{A}+\mathbf{B}) + \frac{(\mathbf{A}+\mathbf{B})^2}{2!} + \ldots, \tag{92}$$

whereas

$$e^{\mathbf{A}}e^{\mathbf{B}} = \left(\mathbf{I}+\mathbf{A}+\frac{\mathbf{A}^2}{2!}+\ldots\right)\left(\mathbf{I}+\mathbf{B}+\frac{\mathbf{B}^2}{2!}+\ldots\right). \tag{93}$$

Hence

$$e^{\mathbf{A}+\mathbf{B}}-e^{\mathbf{A}}e^{\mathbf{B}} = \tfrac{1}{2}(\mathbf{BA}-\mathbf{AB})+ \\ +\text{terms of higher order in } (\mathbf{BA}-\mathbf{AB}). \tag{94}$$

Consequently

$$e^{\mathbf{A}+\mathbf{B}} = e^{\mathbf{A}}e^{\mathbf{B}} \tag{95}$$

only if **A** and **B** commute. When dealing with functions of two matrices analogous results to those of functions of a scalar variable usually hold only if the two matrices commute. For example,

$$\sin(\mathbf{A}+\mathbf{B}) = \sin\mathbf{A}\cos\mathbf{B}+\cos\mathbf{A}\sin\mathbf{B} \tag{96}$$

only if **A** and **B** commute (as can be verified by using the power series expansions (87) and (88)).

We now illustrate how matrix functions may be simplified using the Cayley-Hamilton theorem.

Example 5. To evaluate

$$e^{\mathbf{A}t} \text{ with } \mathbf{A} = \begin{pmatrix} 0 & 1 \\ -1 & 0 \end{pmatrix}, \tag{97}$$

where t is an arbitrary parameter.

Now

$$e^{\mathbf{A}t} = \mathbf{I}+\mathbf{A}t+\frac{\mathbf{A}^2t^2}{2!}+\ldots \tag{98}$$

But the characteristic equation of **A** is

$$f(\lambda) = \lambda^2+1 = 0. \tag{99}$$

Hence, by the Cayley-Hamilton theorem,

$$f(\mathbf{A}) = \mathbf{A}^2+\mathbf{I} = \mathbf{0}. \tag{100}$$

Consequently from (100) we deduce that

$$\mathbf{A}^2 = -\mathbf{I},\quad \mathbf{A}^3 = -\mathbf{A},\quad \mathbf{A}^4 = \mathbf{I},\quad \mathbf{A}^5 = \mathbf{A}, \tag{101}$$

Hence, using (101) in (98),

$$e^{\mathbf{A}t} = \mathbf{I}+\mathbf{A}t-\frac{\mathbf{I}t^2}{2!}-\frac{\mathbf{A}t^3}{3!}+\frac{\mathbf{I}t^4}{4!}+\frac{\mathbf{A}t^5}{5!}-\ldots \tag{102}$$

$$= \mathbf{I}\left(1-\frac{t^2}{2!}+\frac{t^4}{4!}-\ldots\right)+\mathbf{A}\left(t-\frac{t^3}{3!}+\frac{t^5}{5!}-\ldots\right) \tag{103}$$

$$= \mathbf{I}\cos t+\mathbf{A}\sin t. \tag{104}$$

Hence

$$e^{\begin{pmatrix}0 & 1\\ -1 & 0\end{pmatrix}t} = \begin{pmatrix}1 & 0\\ 0 & 1\end{pmatrix}\cos t + \begin{pmatrix}0 & 1\\ -1 & 0\end{pmatrix}\sin t. \tag{105}$$

Example 6. To evaluate

$$e^{\mathbf{A}t} \text{ with } \mathbf{A} = \begin{pmatrix}0 & 0 & 0\\ 1 & 0 & 0\\ 0 & 1 & 0\end{pmatrix}, \tag{106}$$

where t is an arbitrary parameter.

Now by the Cayley-Hamilton theorem

$$f(\mathbf{A}) = \mathbf{A}^3 = \mathbf{0}. \tag{107}$$

Hence

$$\mathbf{A}^r = \mathbf{0}, \quad r \geq 3, \tag{108}$$

and therefore

$$e^{\mathbf{A}t} = \mathbf{I} + \mathbf{A}t + \frac{\mathbf{A}^2 t^2}{2!}. \tag{109}$$

Evaluating $\mathbf{A}^2$ and inserting in (109) we have

$$e^{\mathbf{A}t} = \begin{pmatrix}1 & 0 & 0\\ 0 & 1 & 0\\ 0 & 0 & 1\end{pmatrix} + t\begin{pmatrix}0 & 0 & 0\\ 1 & 0 & 0\\ 0 & 1 & 0\end{pmatrix} + \frac{t^2}{2}\begin{pmatrix}0 & 0 & 0\\ 0 & 0 & 0\\ 1 & 0 & 0\end{pmatrix} \tag{110}$$

$$= \begin{pmatrix}1 & 0 & 0\\ t & 1 & 0\\ \frac{t^2}{2} & t & 1\end{pmatrix}. \tag{111}$$

Example 7. To evaluate

$$e^{\mathbf{A}} \text{ with } \mathbf{A} = \begin{pmatrix}1 & 2\\ 0 & 1\end{pmatrix}. \tag{112}$$

Now it was shown in Example 4 that

$$\mathbf{A}^r = r\mathbf{A} + (1-r)\mathbf{I}. \tag{113}$$

Consequently

$$\frac{\mathbf{A}^r}{r!} = \frac{\mathbf{A}}{(r-1)!} + \frac{1-r}{r!}\mathbf{I}. \tag{114}$$

Hence, using (114),

$$e^{\mathbf{A}} = \sum_{r=0}^{\infty} \frac{\mathbf{A}^r}{r!} = \mathbf{I}+\mathbf{A}\sum_{r=1}^{\infty}\frac{1}{(r-1)!}+\mathbf{I}\sum_{r=1}^{\infty}\frac{1-r}{r!} \tag{115}$$

$$= \mathbf{I}+\mathbf{A}e+\mathbf{I}(e-1-e) \tag{116}$$

$$= \mathbf{A}e$$

$$= \begin{pmatrix} e & 2e \\ 0 & e \end{pmatrix}. \tag{117}$$

Example 8. We have seen earlier in this chapter that the series $1+z+z^2+\ldots+z^r+\ldots$ converges to the function $\dfrac{1}{1-z}$ for $|z| < 1$. Hence the matrix power series

$$\sum_{r=0}^{\infty} \mathbf{A}^r = \mathbf{I}+\mathbf{A}+\mathbf{A}^2+\ldots+\mathbf{A}^r+\ldots = (\mathbf{I}-\mathbf{A})^{-1} \tag{118}$$

provided all the eigenvalues of $\mathbf{A}$ have moduli < 1. The matrix

$$\mathbf{A} = \begin{pmatrix} \frac{1}{2} & 1 \\ 0 & \frac{1}{2} \end{pmatrix} \tag{119}$$

has eigenvalues $\frac{1}{2}$ (twice) satisfying this condition. To evaluate $\mathbf{A}^r$ we write (as in 7.3)

$$\mathbf{A}^r = a_1\mathbf{A}+a_2\mathbf{I}, \tag{120}$$

and, since the two eigenvalues are equal, use (69) to determine the constants a_1 and a_2. Since $\lambda_1 = \frac{1}{2}$, (69) become

$$\left.\begin{aligned} (\tfrac{1}{2})^r &= \tfrac{1}{2}a_1+a_2, \\ r(\tfrac{1}{2})^{r-1} &= a_1, \end{aligned}\right\} \tag{121}$$

from which we find

$$a_1 = r(\tfrac{1}{2})^{r-1}, \qquad a_2 = (\tfrac{1}{2})^r(1-r). \tag{122}$$

Hence, using (120) and (122),

$$\mathbf{A}^r = (\tfrac{1}{2})^{r-1}\left(r\mathbf{A}+\frac{1-r}{2}\mathbf{I}\right). \tag{123}$$

Hence by (118),

$$(\mathbf{I}-\mathbf{A})^{-1} = \sum_{r=0}^{\infty}\mathbf{A}^r = \mathbf{I}+\mathbf{A}\sum_{r=1}^{\infty} r(\tfrac{1}{2})^{r-1}+\mathbf{I}\sum_{r=1}^{\infty}(1-r)(\tfrac{1}{2})^r. \tag{124}$$

Using the fact that the series in the last two terms of (124) are expressible in terms of the geometric series, we find after some simplification that

$$(\mathbf{I}-\mathbf{A})^{-1} = 4\mathbf{A} = \begin{pmatrix} 2 & 4 \\ 0 & 2 \end{pmatrix}. \tag{125}$$

This result is readily checked since

$$\mathbf{I}-\mathbf{A} = \begin{pmatrix} 1 & 0 \\ 0 & 1 \end{pmatrix} - \begin{pmatrix} \frac{1}{2} & 1 \\ 0 & \frac{1}{2} \end{pmatrix} = \begin{pmatrix} \frac{1}{2} & -1 \\ 0 & \frac{1}{2} \end{pmatrix} \tag{126}$$

and

$$\begin{pmatrix} \frac{1}{2} & -1 \\ 0 & \frac{1}{2} \end{pmatrix}\begin{pmatrix} 2 & 4 \\ 0 & 2 \end{pmatrix} = \begin{pmatrix} 1 & 0 \\ 0 & 1 \end{pmatrix} \tag{127}$$

as required.

We have concentrated so far on showing how powers of matrices (and consequently matrix power series such as $e^{\mathbf{A}t}$) may be evaluated with the help of the Cayley-Hamilton theorem. This method is valid for any matrix. However, an alternative method of evaluating matrix power series exists for matrices which are diagonalisable by means of a similarity transformation. For suppose **A** is a matrix such that

$$\mathbf{U}^{-1}\mathbf{A}\mathbf{U} = \mathbf{D}, \qquad \text{(cf. (49) of Chapter 6)}, \tag{128}$$

where **D** is a diagonal matrix whose elements are the eigenvalues of **A** and **U** is the matrix of the eigenvectors of **A**. Then

$$\mathbf{A} = \mathbf{U}\mathbf{D}\mathbf{U}^{-1}, \qquad \mathbf{A}^2 = \mathbf{U}\mathbf{D}\mathbf{U}^{-1}\mathbf{U}\mathbf{D}\mathbf{U}^{-1} = \mathbf{U}\mathbf{D}^2\mathbf{U}^{-1}, \ldots$$
$$\mathbf{A}^r = \mathbf{U}\mathbf{D}^r\mathbf{U}^{-1}, \qquad (r \text{ integral}) \tag{129}$$

Hence for any analytic function $f(z)$ we can write (using (84) and (85))

$$f(\mathbf{A}) = \mathbf{U}\begin{pmatrix} f(\lambda_1) & & & & \\ & f(\lambda_2) & & 0 & \\ & & \cdot & & \\ & & & \cdot & \\ & & & & \cdot \\ & 0 & & & f(\lambda_n) \end{pmatrix}\mathbf{U}^{-1}, \tag{130}$$

where $\lambda_1, \lambda_2, \ldots, \lambda_n$ are the eigenvalues of **A**.

As an example of (130), we have for any diagonalisable matrix **A**

$$e^{\mathbf{A}t} = \mathbf{U}\begin{pmatrix} e^{\lambda_1 t} & & & & \\ & e^{\lambda_2 t} & & 0 & \\ & & \cdot & & \\ & & & \cdot & \\ & & & & \cdot \\ & 0 & & & e^{\lambda_n t} \end{pmatrix}\mathbf{U}^{-1}. \tag{131}$$

7.5 Differentiation and integration of matrices

Suppose **A** is any matrix (not necessarily square) whose elements a_{ik} are at least once differentiable functions of a scalar parameter t. Then the derivative of **A** with respect to t is defined as the matrix whose elements are the derivatives of the elements of **A**. For example, if

$$\mathbf{A} = \begin{pmatrix} \sin t & t^2 \\ 1 & e^{2t} \end{pmatrix} \tag{132}$$

then

$$\frac{d\mathbf{A}}{dt} = \begin{pmatrix} \cos t & 2t \\ 0 & 2e^{2t} \end{pmatrix}. \tag{133}$$

From this definition it follows that if **A** and **B** are any two matrices for which the product **AB** is defined then

$$\frac{d}{dt}(\mathbf{AB}) = \frac{d\mathbf{A}}{dt}\mathbf{B} + \mathbf{A}\frac{d\mathbf{B}}{dt}. \tag{134}$$

Care is necessary, however, in differentiating matrices. For example, it is *not* generally true that

$$\frac{d}{dt}\mathbf{A}^n = n\mathbf{A}^{n-1}\frac{d\mathbf{A}}{dt} \tag{135}$$

as might have been expected. Rather we have to write

$$\frac{d}{dt}\mathbf{A}^n = \frac{d}{dt}(\mathbf{AA}\ldots\mathbf{A})$$

$$= \frac{d\mathbf{A}}{dt}\mathbf{A}^{n-1} + \mathbf{A}\frac{d\mathbf{A}}{dt}\mathbf{A}^{n-2} + \ldots + \mathbf{A}^{n-1}\frac{d\mathbf{A}}{dt} \tag{136}$$

since, in general, **A** and $d\mathbf{A}/dt$ do not commute.

Likewise (provided $\mathbf{A}^{-1}$ exists) $\frac{d}{dt}\mathbf{A}^{-1}$ must be obtained in the following way:

Since $\mathbf{A}\mathbf{A}^{-1} = \mathbf{I}$ (by definition),

$$\frac{d}{dt}(\mathbf{A}\mathbf{A}^{-1}) = \frac{d\mathbf{A}}{dt}\mathbf{A}^{-1} + \mathbf{A}\frac{d\mathbf{A}^{-1}}{dt} = \mathbf{0}. \tag{137}$$

Pre-multiplying (137) by $\mathbf{A}^{-1}$ we have

$$\frac{d\mathbf{A}^{-1}}{dt} = -\mathbf{A}^{-1}\frac{d\mathbf{A}}{dt}\mathbf{A}^{-1}, \tag{138}$$

which gives the expected result $-\mathbf{A}^{-2}\frac{d\mathbf{A}}{dt}$ only when $\mathbf{A}^{-1}$ and $\frac{d\mathbf{A}}{dt}$ commute.

An important result which we use shortly is that for a *constant* matrix $\mathbf{A}$

$$\frac{d}{dt}(e^{\mathbf{A}t}) = \mathbf{A}\, e^{\mathbf{A}t} = e^{\mathbf{A}t}\mathbf{A}. \tag{139}$$

This is readily verified by term-by-term differentiation of the power series expansion of $e^{\mathbf{A}t}$.

Lastly we come to integration. The integral of a matrix $\mathbf{A}$ whose elements are integrable functions of a parameter t (say) is the matrix whose elements are the integrals of the elements of $\mathbf{A}$. Thus if

$$\mathbf{A} = \begin{pmatrix} 1 & \cos t & e^t \\ t & t^2 & t^3 \end{pmatrix} \tag{140}$$

then

$$\int \mathbf{A}\, dt = \begin{pmatrix} t & \sin t & e^t \\ \frac{t^2}{2} & \frac{t^3}{3} & \frac{t^4}{4} \end{pmatrix} + \mathbf{C}, \tag{141}$$

where $\mathbf{C}$ is an arbitrary constant matrix of the same order as $\mathbf{A}$.

The results of this section are of use in the solution of linear differential equations by matrix methods. For suppose we have a set of n linear first order equations in n unknown functions $y_1(t)$, $y_2(t), \ldots, y_n(t)$

$$\frac{dy_i}{dt} = \sum_{j=1}^{o} a_{ij}\, y_j, \qquad (i = 1, 2, \ldots, n), \tag{142}$$

where the a_{ij} are constants, and where the initial values $y_i(0)$ are given.

Writing

$$\mathbf{Y}(t) = \begin{pmatrix} y_1(t) \\ y_2(t) \\ \cdot \\ \cdot \\ \cdot \\ y_n(t) \end{pmatrix} \quad \text{and } \mathbf{A} = \begin{pmatrix} a_{11} & a_{12} & \cdots & a_{1n} \\ a_{21} & a_{22} & \cdots & a_{2n} \\ \cdot & & & \cdot \\ \cdot & & & \cdot \\ \cdot & & & \cdot \\ a_{n1} & & \cdots & a_{nn} \end{pmatrix} \tag{143}$$

(143) may be written in matrix form as

$$\frac{d\mathbf{Y}(t)}{dt} = \mathbf{A}\mathbf{Y}(t), \tag{144}$$

where the column vector $\mathbf{Y}(0)$ is given. Using (139), it is easy to see that the solution of (144) is

$$\mathbf{Y}(t) = e^{\mathbf{A}t}\mathbf{Y}(0). \tag{145}$$

The solution of the set of differential equations is equivalent therefore to finding $e^{\mathbf{A}t}$. This may be done by any of the methods discussed in 7.4. For example, if $\mathbf{A}$ is diagonalisable by a similarity transformation then, using (130),

$$e^{\mathbf{A}t} = \mathbf{U}\, e^{\mathbf{D}t}\, \mathbf{U}^{-1}, \tag{146}$$

where $\mathbf{D}$ is a diagonal matrix with the eigenvalues of $\mathbf{A}$ as elements. Consequently (145) becomes

$$\mathbf{Y}(t) = \mathbf{U}\, e^{\mathbf{D}t}\, \mathbf{U}^{-1}\mathbf{Y}(0). \tag{147}$$

Another approach is to make the transformation

$$\mathbf{W}(t) = \mathbf{U}^{-1}\, \mathbf{Y}(t) \tag{148}$$

in (144), whence

$$\frac{d\mathbf{W}(t)}{dt} = \mathbf{U}^{-1}\mathbf{A}\mathbf{U}\mathbf{W}(t) = \mathbf{D}\mathbf{W}(t). \tag{149}$$

This is a set of uncoupled equations of the type

$$\left.\begin{aligned} \frac{d\omega_1(t)}{dt} &= \lambda_1\, \omega_1(t), \\ \frac{d\omega_2(t)}{dt} &= \lambda_2\, \omega_2(t), \\ \cdot \quad & \quad \cdot \\ \cdot \quad & \quad \cdot \\ \cdot \quad & \quad \cdot \\ \frac{d\omega_n(t)}{dt} &= \lambda_n\, \omega_n(t), \end{aligned}\right\} \tag{150}$$

where the $\omega_i(t)$ are the elements of the column vector $\mathbf{W}(t)$, and $\lambda_1, \lambda_2, \ldots, \lambda_n$ are the eigenvalues of $\mathbf{A}$. Each of these equations may be solved separately and $\mathbf{Y}(t)$ found from the inverse transformation of (148).

PROBLEMS 7

1. Evaluate

$$\begin{pmatrix} 1 & 3 \\ 3 & 1 \end{pmatrix}^{14} \quad \text{and} \quad \begin{pmatrix} 1 & 0 \\ 3 & 1 \end{pmatrix}^{30}.$$

2. Show that

$$\begin{pmatrix} \cos\theta & \sin\theta \\ -\sin\theta & \cos\theta \end{pmatrix}^n = \begin{pmatrix} \cos n\theta & \sin n\theta \\ -\sin n\theta & \cos n\theta \end{pmatrix}.$$

3. Show that the eigenvalues of

$$\mathbf{A} = \begin{pmatrix} 1-p & q \\ p & 1-q \end{pmatrix}$$

are $\lambda_1 = 1$, $\lambda_2 = 1-p-q$. Hence (using the method of 7.3) deduce that

$$\mathbf{A}^r = \frac{1}{p+q}\begin{pmatrix} q & q \\ p & p \end{pmatrix} + \frac{(1-p-q)^r}{p+q}\begin{pmatrix} p & -q \\ -p & q \end{pmatrix}$$

assuming that $p+q \neq 0$.

4. Extend the method of 7.3 to an nth order matrix with distinct eigenvalues $\lambda_1, \lambda_2, \ldots, \lambda_n$ so obtaining the result that

$$\mathbf{A}^r = R(\mathbf{A}) = a_1\mathbf{A}^{n-1} + a_2\mathbf{A}^{n+2} + \ldots + a_{n-1}\mathbf{A} + a_n\mathbf{I},$$

where the constants $a_1, a_2, \ldots, a_n$ are uniquely determined by the n equations

$$\lambda_1^r = R(\lambda_1), \quad \lambda_2^r = R(\lambda_2), \ldots, \lambda_n^r = R(\lambda_n).$$

Hence obtain $\mathbf{A}^8$ for

$$\mathbf{A} = \begin{pmatrix} 1 & 1 & 1 \\ 0 & 2 & 1 \\ -4 & 4 & 3 \end{pmatrix}.$$

5. Suppose $\mathbf{A}$ is an nth order matrix with n repeated eigenvalues $\lambda_1 = \lambda_2 = \ldots = \lambda_n$. Show that

$$\mathbf{A}^r = R(\mathbf{A}) = a_1\mathbf{A}^{n-1} + a_2\mathbf{A}^{n-2} + \ldots + a_{n-1}\mathbf{A} + a_n\mathbf{I},$$

where the constants $a_1, a_2, \ldots, a_n$ are determined by the equations

$$\begin{aligned}
\lambda_1^r &= R(\lambda_1),\\
r\lambda_1^{r-1} &= R^{(1)}(\lambda_1) = (n-1)a_1\lambda_1^{n-2} + \\
&\qquad +(n-2)a_2\lambda_1^{n-3} + \\
&\qquad +\ldots+a_{n-1},\\
r(r-1)\lambda_1^{r-2} &= R^{(2)}(\lambda_1) = (n-1)(n-2)a_1\lambda_1^{n-3} + \\
&\qquad +(n-2)(n-3)\times \\
&\qquad \times a_2\lambda_1^{n-4}+\ldots+ \\
&\qquad +2a_{n-2},\\
&\vdots\\
[r(r-1)\ldots(r-n+1)]\lambda_1^{r-n+1} &= R^{(n-1)}(\lambda_1) = n!\,a_1,
\end{aligned}$$

where

$$R^{(n-1)}(\lambda_1) = \left(\frac{d^{n-1}R(\lambda)}{d\lambda^{n-1}}\right)_{\lambda=\lambda_1}.$$

The matrix

$$\mathbf{A} = \begin{pmatrix} 1 & 0 & 0 & 0 \\ 1 & 1 & 0 & 0 \\ 0 & 1 & 1 & 0 \\ -1 & -1 & 0 & 1 \end{pmatrix}$$

has all four of its eigenvalues equal to unity. Show that

$$\mathbf{A}^r = \begin{pmatrix} 1 & 0 & 0 & 0 \\ r & 1 & 0 & 0 \\ \dfrac{r(r-1)}{2} & r & 1 & 0 \\ \dfrac{-r(r+1)}{2} & -r & 0 & 1 \end{pmatrix}.$$

6. Given that $\mathbf{A}$ has distinct eigenvalues, show by diagonalising $\mathbf{A}$ that the condition that $\mathbf{A}^r \to 0$ as $r \to \infty$, where r is a positive integer is that the moduli of all eigenvalues of $\mathbf{A}$ are less than unity.

7. Given that $e^{\mathbf{A}}$ is diagonalisable by a similarity transformation, show that

$$|e^{\mathbf{A}}| = e^{Tr\mathbf{A}}.$$

Hence deduce that $|e^{\mathbf{A}}| = 1$ when $\mathbf{A}$ is a skew-symmetric matrix.

8. Show that if $\mathbf{A}$ is a real skew-symmetric matrix then $e^{\mathbf{A}}$ is an orthogonal matrix.

9. Show that if $\mathbf{H}$ is a Hermitian matrix then $e^{i\mathbf{H}}$ is a unitary matrix. (This result is of extreme importance in quantum mechanics.)

10. Show that

$$e^{\begin{pmatrix} 1 & 0 \\ 0 & 2 \end{pmatrix}} = \begin{pmatrix} e & 0 \\ 0 & e^2 \end{pmatrix}.$$

11. By letting $\mathbf{Y}^2 = \mathbf{A}$, where $\mathbf{A}$ is a function of a parameter t, obtain an equation for $\frac{d}{dt}(\mathbf{A}^{\frac{1}{2}})$.

12. A function x_r defined for $r = 0, 1, 2, \ldots$ satisfies the second-order linear difference equation

$$x_{r+1} + ax_r + bx_{r-1} = 0 \qquad (r = 1, 2, 3 \ldots),$$

and is subject to the initial conditions

$$x_0 = \alpha, \qquad x_1 = \beta,$$

where α and β are given constants.

By writing $y_{r+1} = x_r$, the second-order difference equation may be written as a pair of first-order difference equations

$$x_{r+1} = -ax_r - by_r,$$

$$y_{r+1} = x_r.$$

Letting

$$\mathbf{E}_r = \begin{pmatrix} x_r \\ y_r \end{pmatrix}$$

this pair of equations may be written in matrix form as

$$\mathbf{E}_{r+1} = \mathbf{A}\mathbf{E}_r,$$

where

$$\mathbf{A} = \begin{pmatrix} -a & -b \\ 1 & 0 \end{pmatrix}.$$

Hence

$$\mathbf{E}_{r+1} = \mathbf{A}\mathbf{E}_r = \mathbf{A}^2\mathbf{E}_{r-1} = \ldots = \mathbf{A}^r\mathbf{E}_1,$$

where

$$\mathbf{E}_1 = \begin{pmatrix} x_1 \\ y_1 \end{pmatrix} = \begin{pmatrix} x_1 \\ x_0 \end{pmatrix} = \begin{pmatrix} \beta \\ \alpha \end{pmatrix}.$$

By evaluating $\mathbf{A}^r$ for the difference equation

$$x_{r+1}-5x_r+6x_{r-1}=0,$$

solve this equation for x_r subject to the initial conditions $x_0 = 1,\ x_1 = 2.$

13. Show that the set of n linear differential equations

$$\frac{dy_i(t)}{dt}=\sum_{j=1}^{n} a_{ij}\, y_j(t)+f_i(t)$$

in the unknowns $y_i(t)$, where a_{ij} are constants and $f_i(t)$ are given functions, may be written in matrix form as

$$\frac{d\mathbf{Y}(t)}{dt}=\mathbf{AY}(t)+\mathbf{F}(t),$$

where

$$\mathbf{Y}(t)=\begin{pmatrix} y_1(t) \\ y_2(t) \\ \cdot \\ \cdot \\ \cdot \\ y_n(t) \end{pmatrix}, \qquad \mathbf{A}=\begin{pmatrix} a_{11} & a_{12} & \cdots & a_{1n} \\ a_{21} & a_{22} & \cdots & a_{2n} \\ \cdot & & & \cdot \\ \cdot & & & \cdot \\ \cdot & & & \cdot \\ a_{n1} & \cdot & \cdot \cdot \cdot & a_{nn} \end{pmatrix}$$

and

$$\mathbf{F}(t)=\begin{pmatrix} f_1(t) \\ f_2(t) \\ \cdot \\ \cdot \\ \cdot \\ f_n(t) \end{pmatrix}.$$

Show that the solution of this matrix equation is

$$\mathbf{Y}(t)=e^{\mathbf{A}t}\mathbf{Y}(0)+\int_0^t e^{\mathbf{A}(t-t')}\mathbf{F}(t')\,dt'.$$

14. Verify that the solution of the matrix equation

$$\frac{d\mathbf{Y}(t)}{dt}=\mathbf{AY}(t)+\mathbf{Y}(t)\mathbf{B},$$

where $\mathbf{A}$ and $\mathbf{B}$ are constant matrices, and where $\mathbf{Y}(0)=\mathbf{C}$ is a constant matrix, is

$$\mathbf{Y}(t)=e^{\mathbf{A}t}\mathbf{C}e^{\mathbf{B}t}.$$

15. Show that a necessary condition that the solution of matrix equation

$$\frac{d\mathbf{Y}(t)}{dt} = \mathbf{A}(t), \qquad \mathbf{Y}(0) = \mathbf{B},$$

where **A** and **B** are constant matrices and **A** has distinct eigenvalues, tends to zero as $t \to \infty$ is that all the eigenvalues of **A** have negative real parts. (Hint: use the representation of (131).)

CHAPTER 8

Group Theory

8.1 Introduction

We mentioned in Chapter 1, 1.1 that set theory led naturally on into group theory. Now in this final chapter, having dealt with sets and matrices in the earlier chapters, we discuss what is required of a set of elements in order that it should be a group, and, in addition, show the way in which matrices play an important part in what is called group representation theory. Group theory is an important subject principally in the fields of theoretical physics and chemistry and, within this context, finds numerous applications to the quantum mechanics of atoms, molecules and nuclei, solid state theory, crystal structure, as well as to elementary particle theory and relativity.

Group theory is the formal mathematical way of dealing with the symmetries (if any) of a system or structure, and its importance lies therefore in simplifying the mathematical description of the system in virtue of any symmetries it may have. Some elementary examples of groups are given in 8.3. However, most of the applications to genuine physical problems require an extensive knowledge of the subject to which group theory is being applied (e.g. quantum mechanics, crystal structure). Rather than attempt to give the necessary background to these subjects and then demonstrate the applications of group theory, it was felt better to provide the basic language of group theory, leaving it to the reader to apply it to his particular subject. To this end, the list of further reading matter at the end of the book provides a fairly wide selection of books covering most of the fields mentioned here.

8.2 Group axioms

A set G (finite or infinite) of elements $a, b, c \dots$ is said to form a group if there exists a rule for combining any two elements to form their ' product ' ab, say, such that the following four axioms are satisfied.

(i) For every $a,b \in G$ (using the set notation of Chapter 1, 1.2), $ab \in G$. In other words, every ' product ' of two elements (ab being considered as different, in general, from ba) and every ' square ' (aa) are to be elements of G.

If this axiom is satisfied the set is said to be closed under multiplication.

(ii) For every $a,b,c \in G$,

$$(ab)c = a(bc).$$

This is the associative law for group ' products '.

(iii) The set G contains a unit (null, or neutral) element e such that for all $a \in G$

$$ae = ea = a. \tag{1}$$

(iv) For every $a \in G$ there exists an element a^{-1} of G called the inverse of a such that

$$aa^{-1} = a^{-1}a = e. \tag{2}$$

The word ' product ' used here is to be understood within the context of the rule of combination. For example, if the elements are to be combined under multiplication then their ' products ' are obtained by multiplying any two elements together. If, however, the rule of combination is addition then the ' product ' of any two elements is their sum.

A group is called Abelian (or commutative) if for every pair of elements $a,b \in G$

$$ab = ba. \tag{3}$$

Finally, any finite set of elements satisfying the four group axioms is said to form a finite group, the *order* of the group being equal to the number of elements in the set. If the group does not have a finite number of elements it is called an infinite group.

Examples of these various types of groups will be given in the next section. However, before doing this it might reasonably be asked whether the unit element in (iii) is necessarily unique. To show that this is so we suppose that e and e' are two unit elements of G. Then by (1)

$$ae = ea = a, \tag{4}$$

and likewise

$$ae' = e'a = a. \tag{5}$$

Letting $a = e$ in (4) and (5), we have

$$e^2 = e = ee' = e'. \tag{6}$$

Similarly, letting $a = e'$ in (4) and (5),

$$e'^2 = e' = e'e = ee'. \tag{7}$$

Hence, comparing (6) and (7), it follows that

$$e = e'. \tag{8}$$

The unit element therefore is unique.

Similarly, it may be proved that the inverse of each group element is unique (see Problem 1 at the end of the chapter).

8.3 Examples of groups

Example 1. The set S_1 of all integers (positive, negative and zero) forms an infinite group under addition. To verify this we note that the group axiom (i) is satisfied since the sum of any two integers (and the sum of any integer with itself) is always another integer. Similarly, (ii) is satisfied since the associative law of addition $a+(b+c) = (a+b)+c$ is true for integers. The unit element must be taken as 0, since the addition of 0 to any integer does not alter it; consequently (iii) is satisfied. Finally, (iv) is satisfied since, if the inverse of an integer is defined as its negative, then

$$a+(-a) = 0.$$

The group is Abelian since $a+b = b+a$.

We note here that the same set does not form a group under multiplication since the inverses of integers are not integers; (iv) therefore cannot be satisfied.

Example 2. The set S_2 of all rational numbers p/q ($q \neq 0$) forms a group under addition. Here the unit element is 0 (i.e. $p = 0$) and the inverse of a given number is its negative. Again this is an example of an infinite Abelian group.

Example 3. The set S_3 of all complex numbers $z = x+iy$ forms an infinite Abelian group under addition. Here $z = 0$ is the unit element, and $-z$ is the inverse of z.

We notice that the set of elements of Example 1 is a subset of the set of elements of Example 2. Likewise, the set of elements of Example 2 is a subset of the set of elements of Example 3. Hence

$$S_1 \subset S_2 \subset S_3 \qquad \text{(and hence } S_1 \subset S_3\text{)}. \tag{9}$$

Since each of these sets forms a group under the same rule of combination, we say that S_1 is a subgroup of S_2 and that S_2 is a subgroup of S_3. Accordingly, S_1 is a subgroup of S_3.

Subgroups will be dealt with in more detail in 8.9. For the moment, however, we remark that every group G (say) has two trivial or improper subgroups, namely G itself and the group containing only one element – the unit element.

Example 4. It may easily be verified that the set of all rational numbers, the set of all real numbers and the set of all complex numbers, with 0 excluded in each case, form infinite Abelian groups under multiplication. For example, if we take a rational number p/q ($p \neq 0$, $q \neq 0$) then (i) is satisfied since the product (in the ordinary sense) of two rational numbers is another rational number. Axiom (ii) is clearly satisfied since multiplication of numbers is associative. The unit element is $1/1 = 1$, and if the inverse of p/q is taken as q/p then (iv) is satisfied since $(p/q)(q/p) = 1$.

Example 5. Consider now the rotations of a line about the z-axis through angles $\pi/2$, π, $3\pi/2$ and 2π in the xy-plane (see Fig. 8.1).

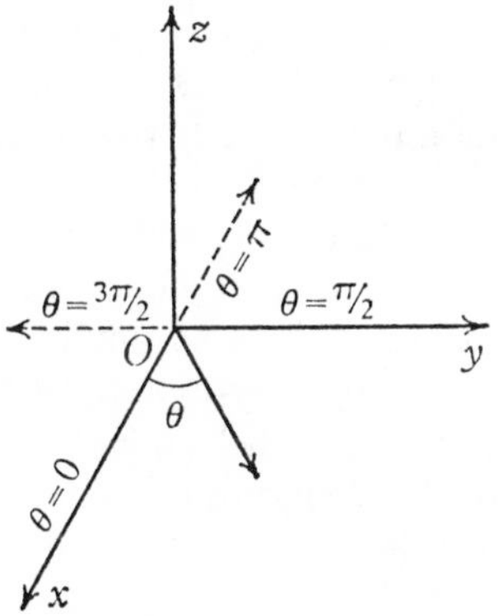

Fig. 8.1

This is a finite set of order 4 in that it contains four elements, namely the four rotations through angles of $\pi/2$, π and so on. We now show that this set of elements forms a group under composition of rotations. It is clear that if we perform the operation of rotating the line through $\pi/2$ from the $\theta = 0$ position and then follow it by a further rotation of π we reach the $\theta = 3\pi/2$ position. This position could have been reached by performing one basic rotation of $3\pi/2$. Likewise the composition of any two basic rotations leads to another basic rotation. Consequently the group axiom (i) is satisfied. The associative law (ii) is satisfied since the order in which successive rotations are performed is immaterial, e.g.

$$\pi + \left(\frac{3\pi}{2} - 2\pi\right) = \left(\pi + \frac{3\pi}{2}\right) - 2\pi$$

The element $\theta = 2\pi(\equiv 0°)$ corresponds to the unit element, since a rotation of the line through 2π brings it back to its initial position. Hence (iii) is satisfied. Finally, (iv) is satisfied if the inverse of any basic rotation is defined as a rotation of the same magnitude but in the opposite direction.

Example 6. The set of four numbers $1, i, -1, -i$ forms a group of order 4 under multiplication. Group property (i) is clearly satisfied since the product of any two elements (and the squares of each element) are elements of the set (e.g. $1i = i$, $i(-i) = 1$, $i^2 = -1$, $(-i)^2 = -1$, etc.). The associative law (ii) also holds for the multiplication of numbers. The unit element e is taken as the number 1. Finally, if the inverse of every element is taken as its reciprocal (e.g. $1/i = -i$, $1/-1 = -1$, etc.) then group property (iv) is satisfied.

Example 7. It should now be clear from the earlier work on matrices that they possess properties such that the set of all square non-singular matrices of a fixed order forms an infinite group under matrix multiplication, the unit matrix corresponding to the unit element of the group. This group is non-Abelian. However, finite sets of non-singular matrices may also form groups. As an example of a finite group of matrices which is Abelian we give the matrices

$$\begin{pmatrix} 1 & 0 \\ 0 & 1 \end{pmatrix}, \quad \begin{pmatrix} 0 & 1 \\ -1 & 0 \end{pmatrix}, \quad \begin{pmatrix} -1 & 0 \\ 0 & -1 \end{pmatrix}, \quad \begin{pmatrix} 0 & -1 \\ 1 & 0 \end{pmatrix}, \tag{10}$$

which form a group of order 4 under matrix multiplication.

Sets of matrices which form groups with respect to matrix multiplication are usually called matrix groups, and are of extreme importance in the theory of group representations (see 8.10).

8.4 Cyclic groups

A group whose elements can all be expressed as powers of a single element is called a cyclic group. The structure of the group is such that the set of elements

$$e, a, a^2, \ldots, a^{n-1}, \tag{11}$$

where n is the smallest integer for which

$$a^n = e, \tag{12}$$

forms the cyclic group of order n generated by the element a. The first three group axioms are easily verified by a direct inspection of (11) and (12); we now verify that each element possesses an inverse element in the set. To do this we simply note that, since

$$a^r a^{n-r} = a^n = e, \tag{13}$$

the inverse of a^r is a^{n-r} which is an element of the set. Hence group axioms (iv) is satisfied.

Cyclic groups are necessarily Abelian since $a^2 a = aa^2$, etc.

We now give some examples of cyclic groups.

Example 8. Suppose PQR is an equilateral triangle (see Fig. 8.2).

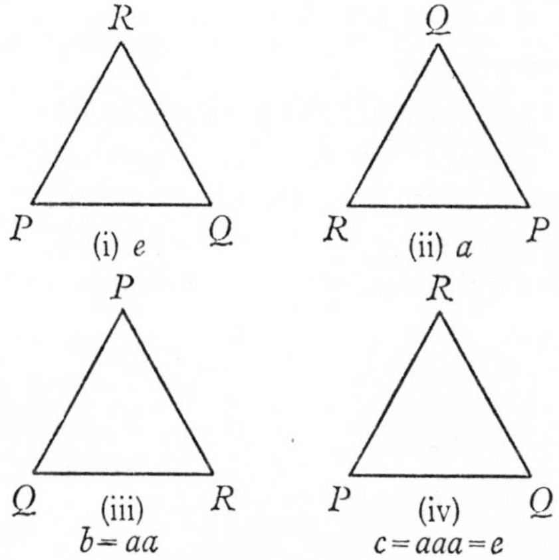

Fig. 8.2

Consider the rotations of PQR in its plane which bring it into coincidence with itself. These rotations may be represented as follows:

e ($\equiv 0°$) leaves PQR unchanged (see Fig. 8.2(i)),

$a\ (\equiv 2\pi/3)$ sends $P\to Q$, $Q\to R$, $R\to P$ (see Fig. 8.2(ii)),
$b = aa\ (\equiv 4\pi/3)$ sends $P\to R$, $Q\to P$, $R\to Q$ (see Fig. 8.2(iii)),
$c = aaa\ (\equiv 2\pi)$ brings P back to P, Q back to Q, and R back to R (see Fig. 8.2(iv)).

Clearly $c = aaa = a^3 = e$. Hence the possible rotations form a set of three elements e, a and b, or equivalently

$$e,\ a,\ a^2(a^3 = e). \tag{14}$$

This set forms a cyclic group of order 3 under composition of rotations. For example, the rotation b is equivalent to the rotation a twice over (i.e. $4\pi/3 = 2\pi/3+2\pi/3$). Similarly, the inverse of a is a^2 since the rotation which undoes the work of $a\ (\equiv 2\pi/3)$ is a further rotation $b\ (\equiv 4\pi/3)$.

Example 9. The set of elements

$$1,\ a,\ a^2,\ldots,\ a^{n-1}, \tag{15}$$

where

$$a = \exp(2\pi i/n), \tag{16}$$

forms a cyclic group of order n under multiplication. Again $a^n = 1\ (\equiv e)$ as required by (12).

8.5 Group tables

A group of order n clearly has n^2 products. These products may be arranged in a square array called a group multiplication table. As a particular example we take the group of order 4 of Example 6, where the elements e, a, b, c are the numbers $1, i, -1, -i$ respectively. The group multiplication table then takes the form

	e	a	b	c
	e	a	b	c
a	a	b	c	e
b	b	c	e	a
c	c	e	a	b

or equivalently

	1	i	-1	$-i$
1	1	i	-1	$-i$
i	i	-1	$-i$	1
-1	-1	$-i$	1	i
$-i$	$-i$	1	i	-1

Table 1.

from which the product ab (say) may be read off as the element common to the row marked a and the column marked b (in that order). Since the group is Abelian ($ab = ba$, etc.) the multiplication table is symmetrical about its leading diagonal. Conversely, if a group multiplication table is symmetrical about its leading diagonal it must arise from an Abelian group.

In a group multiplication table each element occurs once only in each row, and once only in each column. For if the elements of the group are a_i ($i = 1, 2, \ldots, n$) and if two entries in a row or column are the same then $a_i a_j = a_i a_k$. This gives $a_j = a_i^{-1} a_i a_k = a_k$ which is not the case.

We now give another example of a group and its multiplication table which will be of interest again in later sections of this chapter.

Example 10. Consider all the rotations which send an equilateral triangle into itself. (This is not the same problem as in Example 8, where only rotations in the plane were allowed.) Now let PQR be an equilateral triangle with centre O (see Fig. 8.3), and let OA, OB and

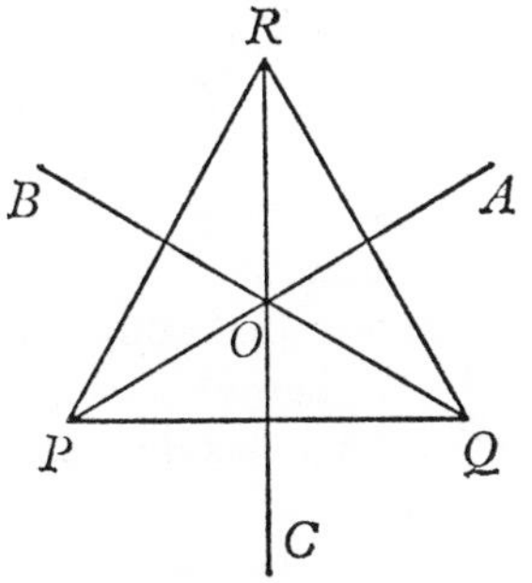

Fig. 8.3

OC be a set of rotation axes fixed in space and passing through the three vertices of the triangle (these axes are left unchanged as the triangle is rotated).

The operations which bring the triangle into coincidence with itself may now be described as follows:

- e: the identity element (leave the triangle as it is).
- a: an anti-clockwise rotation of $2\pi/3$ in the plane of the triangle so that $P \to Q$, $Q \to R$, $R \to P$.

b: an anti-clockwise rotation of $4\pi/3$ in the plane of the triangle so that $P \to R$, $Q \to P$, $R \to Q$.

μ: a rotation of the triangle through π about the OA axis.

ν: a rotation of the triangle through π about the OB axis.

σ: a rotation of the triangle through π about the OC axis.

As in Example 8, we see that $b = a^2$. All the products of the elements may now be worked out from first principles, and it is easily verified that this set of six operations forms a group. For example, μb means *first* consider the effect of b and *then* the effect of μ. Now b sends

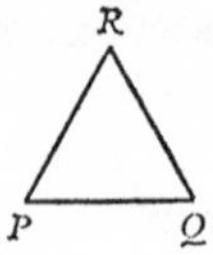

into

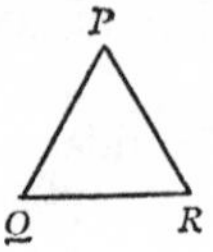

and the effect of μ on this is to send

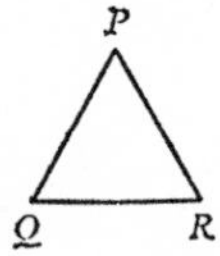

into

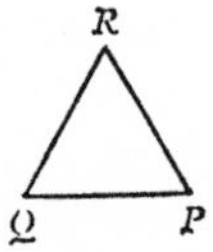

This configuration is the same as that obtained by rotating the original configuration through π about the OC axis. Hence $\mu b = \sigma$. Similarly $b\mu = \nu$, which shows that, since $\mu b \neq b\mu$, the group is non-Abelian.

The group multiplication table has the following form:

	e	a	b	μ	ν	σ
e	e	a	b	μ	ν	σ
a	a	b	e	σ	μ	ν
b	b	e	a	ν	σ	μ
μ	μ	ν	σ	e	a	b
ν	ν	σ	μ	b	e	a
σ	σ	μ	ν	a	b	e

Table 2.

From this table the inverses of the six elements may easily be read off. For example, $b^{-1} = a$, $\mu^{-1} = \mu$, etc. Furthermore, we notice that the set of elements $\{e, a, b\}$ forms a subgroup (see 8.3 and 8.9) of order 3 and that, since $b = a^2$, this subgroup is cyclic with the element a as the generator (see 8.4). Three other subgroups – each of order 2 – exist, namely $\{e, \mu\}$, $\{e, \nu\}$ and $\{e, \sigma\}$.

8.6 Isomorphic groups

Two groups with the same multiplication table are called isomorphic. In more formal language two groups G and G' with elements $a, b, c \ldots$ and $a', b', c' \ldots$ respectively are said to be isomorphic if a one-to-one correspondence exists between all their elements such that $ab = c$ implies $a'b' = c'$, etc., and vice versa. The elements of the two groups may, however (and, in general, do), represent completely different mathematical entities.

The isomorphism of groups is a special instance of the homomorphism of groups. For, whereas isomorphism requires a one-to-one correspondence between elements, homomorphism allows a one-to-many correspondence. However, we shall not discuss this concept further here.

We now give two examples of the isomorphism of groups.

Example 11. Consider the group G of Example 6. This consists of the four elements

$$e = 1, \quad a = i, \quad b = -1, \quad c = -i, \tag{17}$$

with ordinary multiplication as the rule of combination. The group multiplication table is shown in Table 1.

Now let G' be the matrix group of Example 7, with elements

$$e' = \begin{pmatrix} 1 & 0 \\ 0 & 1 \end{pmatrix}, \quad a' = \begin{pmatrix} 0 & 1 \\ -1 & 0 \end{pmatrix}, \quad b' = \begin{pmatrix} -1 & 0 \\ 0 & -1 \end{pmatrix}, \quad c' = \begin{pmatrix} 0 & -1 \\ 1 & 0 \end{pmatrix}. \tag{18}$$

It is easily found that the group multiplication table of this group is

	e'	a'	b'	c'
e'	e'	a'	b'	c'
a'	a'	b'	c'	e'
b'	b'	c'	e'	a'
c'	c'	e'	a'	b'

Table 3.

Comparing Tables 1 and 3 we see that they have precisely the same structure. Accordingly, the groups G and G' are isomorphic.

Example 12. The matrix group G' of order 6 with elements

$$e' = \begin{pmatrix} 1 & 0 \\ 0 & 1 \end{pmatrix}, \quad a' = \begin{pmatrix} -\frac{1}{2} & \frac{\sqrt{3}}{2} \\ -\frac{\sqrt{3}}{2} & -\frac{1}{2} \end{pmatrix}, \quad b' = \begin{pmatrix} -\frac{1}{2} & -\frac{\sqrt{3}}{2} \\ \frac{\sqrt{3}}{2} & -\frac{1}{2} \end{pmatrix} \tag{19}$$

$$\mu' = \begin{pmatrix} 1 & 0 \\ 0 & -1 \end{pmatrix}, \quad \nu' = \begin{pmatrix} -\frac{1}{2} & \frac{\sqrt{3}}{2} \\ \frac{\sqrt{3}}{2} & \frac{1}{2} \end{pmatrix}, \quad \sigma' = \begin{pmatrix} -\frac{1}{2} & -\frac{\sqrt{3}}{2} \\ -\frac{\sqrt{3}}{2} & \frac{1}{2} \end{pmatrix}$$

is isomorphic with the group G of Example 10, as may be verified by constructing the multiplication table of G' and comparing with Table 2. For example,

$$\mu' b' = \begin{pmatrix} 1 & 0 \\ 0 & -1 \end{pmatrix} \begin{pmatrix} -\frac{1}{2} & -\frac{\sqrt{3}}{2} \\ \frac{\sqrt{3}}{2} & -\frac{1}{2} \end{pmatrix} = \begin{pmatrix} -\frac{1}{2} & -\frac{\sqrt{3}}{2} \\ -\frac{\sqrt{3}}{2} & \frac{1}{2} \end{pmatrix} = \sigma', \tag{20}$$

and so on.

8.7 Permutations: the symmetric group

Suppose we have a set of n distinct objects labelled, for convenience, $1, 2, \ldots, n$. The operation of replacing 1 by a_1, 2 by $a_2, \ldots, n$ by a_n to give some arrangement $a_1 a_2 \ldots a_n$ of the same n objects is called a permutation P and is denoted by the symbol

$$P = \begin{pmatrix} 1 & 2 & 3 & \ldots & n \\ a_1 & a_2 & a_3 & \ldots & a_n \end{pmatrix}, \tag{21}$$

indicating that each element in the first row is to be replaced by the element directly below it in the second row. The order in which the columns of the permutation symbol (21) are placed is irrelevant and we may just as well write

$$P = \begin{pmatrix} 2 & 1 & n & \ldots & 3 \\ a_2 & a_1 & a_n & \ldots & a_3 \end{pmatrix} = \begin{pmatrix} 1 & 3 & 2 & \ldots & n \\ a_1 & a_3 & a_2 & \ldots & a_n \end{pmatrix}, \tag{22}$$

and so on.

For n objects there are $n!$ arrangements or permutations, each of which may be written in the form (21).

To be more explicit we deal here with the set of six ($= 3!$) permutations of three objects. These permutations are

$$\left.\begin{aligned} P_1 &= \begin{pmatrix} 1 & 2 & 3 \\ 1 & 2 & 3 \end{pmatrix}, & P_2 &= \begin{pmatrix} 1 & 2 & 3 \\ 3 & 1 & 2 \end{pmatrix}, \\ P_3 &= \begin{pmatrix} 1 & 2 & 3 \\ 2 & 3 & 1 \end{pmatrix}, & P_4 &= \begin{pmatrix} 1 & 2 & 3 \\ 2 & 1 & 3 \end{pmatrix}, \\ P_5 &= \begin{pmatrix} 1 & 2 & 3 \\ 3 & 2 & 1 \end{pmatrix}, & P_6 &= \begin{pmatrix} 1 & 2 & 3 \\ 1 & 3 & 2 \end{pmatrix}. \end{aligned}\right\} \quad (23)$$

Now the product of two permutations P_iP_j ($i,j = 1, 2, \ldots, 6$) is defined as the permutation obtained by first performing P_j and then P_i. (This convention is consistent with that used for operators, although when dealing with permutations the opposite convention – P_i first, then P_j – is adopted in many texts.) For example, P_6P_2 means first perform P_2 and then P_6. To evaluate the result we see that by P_2 1 is replaced by 3, and by P_6 3 is replaced by 2. Hence, by P_6P_2, 1 is replaced by 2.

Similarly by P_2 2 is replaced by 1, and by P_4 1 is replaced by 1. Hence, by P_6P_2, 2 is replaced by 1. Finally, we find

$$P_6P_2 = \begin{pmatrix} 1 & 2 & 3 \\ 1 & 3 & 2 \end{pmatrix}\begin{pmatrix} 1 & 2 & 3 \\ 3 & 1 & 2 \end{pmatrix} = \begin{pmatrix} 1 & 2 & 3 \\ 2 & 1 & 3 \end{pmatrix} = P_4. \quad (24)$$

Other products may be obtained in the same way.

Included in the set of six permutations is the one which leaves the original arrangement unaltered, namely P_1. This permutation is called the identity permutation.

Lastly, to every permutation P_i there exists another permutation P_i^{-1} called the inverse of P_i which undoes the work of P_i. For example the inverse of

$$P_2 = \begin{pmatrix} 1 & 2 & 3 \\ 3 & 1 & 2 \end{pmatrix} \quad (25)$$

is

$$P_2^{-1} = \begin{pmatrix} 3 & 1 & 2 \\ 1 & 2 & 3 \end{pmatrix} = \begin{pmatrix} 1 & 2 & 3 \\ 2 & 3 & 1 \end{pmatrix} = P_3, \quad (26)$$

since

$$P_2P_2^{-1} = P_2P_3 = P_1 \text{ (the identity permutation).} \quad (27)$$

Now from (24) it is seen that the product P_6P_2 ($= P_4$) is an element of the set of six basic permutations (23). Likewise by (26) the inverse of P_2 ($= P_3$) is also an element of the set. In fact, it may easily be verified that all products and inverses are elements of the set of permutations and that accordingly the permutations $P_1, P_2, \ldots, P_6$ form a (non-Abelian) group of order 6. This group is called the symmetric group, and its multiplication table is shown in Table 4.

	P_1	P_2	P_3	P_4	P_5	P_6
P_1	P_1	P_2	P_3	P_4	P_5	P_6
P_2	P_2	P_3	P_1	P_6	P_4	P_5
P_3	P_3	P_1	P_2	P_5	P_6	P_4
P_4	P_4	P_5	P_6	P_1	P_2	P_3
P_5	P_5	P_6	P_4	P_3	P_1	P_2
P_6	P_6	P_4	P_5	P_2	P_3	P_1

Table 4.

A comparison of Tables 4 and 2 shows that they have the same structure. The symmetric group of order 6 is therefore isomorphic with the group of operations which bring an equilateral triangle into coincidence with itself (see Example 10). The correspondence between the elements is

$$e \leftrightarrow P_1, \quad a \leftrightarrow P_2, \quad b \leftrightarrow P_3, \quad \mu \leftrightarrow P_4, \quad \nu \leftrightarrow P_5, \quad \sigma \leftrightarrow P_6. \tag{28}$$

Furthermore, since the group of matrices (19) is isomorphic with the group of Example 10, this matrix group must also be isomorphic with the symmetric group of order 6. Another isomorphism is obtained from the correspondence

$$P_1 \leftrightarrow S_1, \quad P_2 \leftrightarrow S_2, \quad P_3 \leftrightarrow S_3, \quad P_4 \leftrightarrow S_4, \quad P_5 \leftrightarrow S_5, \quad P_6 \leftrightarrow S_6, \tag{29}$$

where

$$\left.\begin{aligned}
S_1 &= \begin{pmatrix} 1 & 0 & 0 \\ 0 & 1 & 0 \\ 0 & 0 & 1 \end{pmatrix}, & S_2 &= \begin{pmatrix} 0 & 1 & 0 \\ 0 & 0 & 1 \\ 1 & 0 & 0 \end{pmatrix}, \\
S_3 &= \begin{pmatrix} 0 & 0 & 1 \\ 1 & 0 & 0 \\ 0 & 1 & 0 \end{pmatrix}, & S_4 &= \begin{pmatrix} 1 & 0 & 0 \\ 0 & 0 & 1 \\ 0 & 1 & 0 \end{pmatrix}, \\
S_5 &= \begin{pmatrix} 0 & 1 & 0 \\ 1 & 0 & 0 \\ 0 & 0 & 1 \end{pmatrix}, & S_6 &= \begin{pmatrix} 0 & 0 & 1 \\ 0 & 1 & 0 \\ 1 & 0 & 0 \end{pmatrix}.
\end{aligned}\right\} \tag{30}$$

Although we have dealt specifically here with the permutations of three objects, it is clear that the set of permutations of n objects forms a group – the symmetric group – of order $n!$. As we shall see in the next section this group has an important place in the theory of finite groups as a whole.

We note here in passing that the matrix S_j ($j = 1, 2, \ldots, 6$) of (30) is the unit matrix in which the rows have been subjected to the permutation P_j. This gives a general rule for writing down the $(n \times n)$ matrices associated with the permutation of n objects.

8.8 Cayley's theorem

This theorem states that every finite group is isomorphic with a suitable group of permutations. To prove this important result we let G be a group of order n with elements

$$a_1, a_2, \ldots, a_n. \tag{31}$$

Now choose any one of these elements, say a_i, and form the products (in the group sense)

$$a_i a_1, a_i a_2, \ldots, a_i a_n. \tag{32}$$

These products are again just the n distinct elements of G and consequently form a rearrangement of (31).

Let

$$P_i = \begin{pmatrix} a_1 & a_2 & \ldots & a_n \\ a_i a_1 & a_i a_2 & \ldots & a_i a_n \end{pmatrix} \tag{33}$$

be a permutation associated with the element a_i. Then when a_i is chosen to be the unit element of the group P_i becomes the identity permutation.

Furthermore, if

$$P_j = \begin{pmatrix} a_1 & a_2 & \ldots & a_n \\ a_j a_1 & a_j a_2 & \ldots & a_j a_n \end{pmatrix} \tag{34}$$

is the permutation associated with the element a_j, and a_i and a_j are chosen to be different, then P_i and P_j are different permutations.

Finally, taking the product $P_i P_j$ we have

$$P_i P_j = \begin{pmatrix} a_1 & a_2 & \ldots & a_n \\ a_i a_1 & a_i a_2 & \ldots & a_i a_n \end{pmatrix} \begin{pmatrix} a_1 & a_2 & \ldots & a_n \\ a_j a_1 & a_j a_2 & \ldots & a_j a_n \end{pmatrix} \tag{35}$$

$$= \begin{pmatrix} a_1 & a_2 & \ldots & a_n \\ a_i a_j a_1 & a_i a_j a_2 & \ldots & a_i a_j a_n \end{pmatrix}, \tag{36}$$

which is just the permutation corresponding to the element $a_i a_j$ of the group G.

From these results it is clear that a one-to-one correspondence exists between the elements $a_1, a_2, \ldots, a_n$ of G and the permutations $P_1, P_2, \ldots, P_n$, and that these n permutations themselves form a group. This group H (say) is a subgroup of order n of the symmetric group of order $n!$ which contains all $n!$ permutations of $a_1, a_2, \ldots, a_n$.

Cayley's theorem highlights the important position of permutation groups in the study of finite groups. Quite apart from this, however, permutation groups are of importance in quantum mechanics where, owing to the identity of elementary particles of a given type (all electrons are identical!), various quantities must be invariant under interchange or permutation of the particles. Further details of the consequences of this invariance property may be found in almost any book dealing with the applications of group theory to quantum mechanics.

8.9 Subgroups and cosets

The idea of a subgroup has been met in earlier sections of this chapter. We now prove that the order of a subgroup is a factor of the order of the group from which the subgroup is derived.

Let G be a group of order n with elements

$$a_1, a_2, \ldots, a_n \tag{37}$$

where, for convenience, we associate a_1 with the unit element e. Suppose now H is a subgroup of G of order m with elements

$$b_1, b_2, \ldots, b_m. \tag{38}$$

Again we let $b_1 = e$ (since, being a group H must contain the unit element). We now take some element a_k (say) of G which is *not* in H, and consider the set of m elements

$$b_1 a_k, b_2 a_k, \ldots, b_m a_k. \tag{39}$$

This collection of elements is called the right-coset of H with respect to a_k and is denoted more compactly by Ha_k. (The term ' right ' is used to signify that the ' products ' are obtained by putting the a_k on the right-hand sides of the b_i.) We see that Ha_k consists of m *different* elements since $b_i a_k = b_j a_k$ implies

$$b_i = b_i a_k a_k^{-1} = b_j a_k a_k^{-1} = b_j$$

which is not so.

This right-coset of H does not form a group. For if it did it would contain the unit element ($a_1 = b_1 = e$) which would require for some b_j that $b_j a_k = e$ or, equivalently, $a_k = b_j^{-1}$. This requires a_k to be an element of H, which is contrary to assumption. Indeed, the right-coset does not contain any element in common with H. For supposing it contains some element b_j (say) of H. Then for some b_i we must have

$$b_i a_k = b_j. \tag{40}$$

But this requires that

$$a_k = b_i^{-1} b_j, \tag{41}$$

which, in turn, requires that a_k be a member of H. Again this is contrary to assumption.

Moreover, if Ha_k and Ha_l have an element in common then they are identical. For if $b_i a_k = b_j a_l$ then for any $b_p a_k$ in Ha_k we have

$$b_p a_k = (b_p b_i^{-1}) b_i a_k = (b_p b_i^{-1}) b_j a_l = b_s a_l, \tag{42}$$

where $b_s = b_p b_i^{-1} b_j$. Since $b_s a_l$ is an element of Ha_l we have $Ha_k = Ha_l$. Now every element a_k in G but not in H belongs to some coset Ha_k. Thus G falls into the union of H and a number of non-overlapping cosets, each having m different elements. The order of G is therefore divisible by m. Hence the order of a subgroup H of G is a factor of the order of G.

This result is well illustrated by the group of order 6 of Example 10 where the proper subgroups are of orders 2 and 3. The improper subgroups – namely, the unit element and the group itself – have orders 1 and 6 respectively, both orders again being factors of 6.

Finally, we remark that by forming the products $a_k H$ we obtain the left-cosets of H with respect to a_k. It may be shown that using left-cosets similar arguments to those used here for right-cosets again lead to (42).

8.10 Some remarks on representations

In some previous sections of this chapter (see 8.6 and 8.7) we have seen examples of groups having isomorphisms with matrix groups. In these examples every element of a group G corresponds to a distinct square matrix of a matrix group G' (say). When this is the case G' is called a faithful (or true) representation of G. Suppose G has elements $a, b, \ldots$. Let $\mathbf{\Gamma}(a)$ be the square matrix corresponding to the element a, $\mathbf{\Gamma}(b)$ be the square matrix corresponding to the element b, and so on. Then if

$$\mathbf{\Gamma}(a)\mathbf{\Gamma}(b) = \mathbf{\Gamma}(ab) \tag{43}$$

and

$$\mathbf{\Gamma}(e) = \mathbf{I} \text{ (the unit matrix)} \tag{44}$$

(e being the unit element) the matrices satisfy the group axioms. The order of $\mathbf{\Gamma}$ is called the dimensionality of the representation. For example, the matrices of (18) form a 2-dimensional representation of the group G of Example 6. Likewise, the matrices (19) form a 2-dimensional representation of the group of Example 10. Similarly, the matrices (30) form a 3-dimensional representation of the permutation group of order 6.

Now suppose each matrix of the representation G' of G is transformed by a similarity transformation (see Chapter 6) into $\mathbf{\Gamma}'(a)$, where

$$\mathbf{\Gamma}'(a) = \mathbf{U}^{-1}\mathbf{\Gamma}(a)\mathbf{U}, \tag{45}$$

$\mathbf{U}$ being a non-singular matrix. Then

$$\mathbf{\Gamma}'(a)\mathbf{\Gamma}'(b) = \mathbf{U}^{-1}\mathbf{\Gamma}(a)\mathbf{U}\mathbf{U}^{-1}\mathbf{\Gamma}(b)\mathbf{U} \tag{46}$$

$$= \mathbf{U}^{-1}\mathbf{\Gamma}(a)\mathbf{\Gamma}(b)\mathbf{U} \tag{47}$$

$$= \mathbf{U}^{-1}\mathbf{\Gamma}(ab)\mathbf{U} \text{ (using (43))} \tag{48}$$

$$= \mathbf{\Gamma}'(ab). \tag{49}$$

Hence the group properties still hold for the transformed matrices and so they too form a true representation of G. In general, representations related in this way are regarded as being equivalent, although, of course, the forms of the individual matrices will be quite different in equivalent representations. With this freedom in the choice of the forms of the matrices it is important to look for some quantity which is an invariant for a given representation. This is found in considering the traces of the matrices forming a representation for, as we have seen in Chapter 6, 6.2, the trace of a matrix is invariant under a similarity transformation. The traces of the matrices forming a representation are called the characters of the representation and are invariant under the transformation (45). Characters play an important part in the theory of group representations and are of importance in many of the group theory applications to quantum mechanics. This topic, and others related to whether a given representation of a group can be reduced to one of smaller dimension are, however, beyond the scope of this book.

PROBLEMS 8

1. Show that the inverse of each element of a group is unique.

2. Show that the identity operation and the rotations through π about any one of three mutually perpendicular intersecting lines form a group of order 4, and obtain its multiplication table.

3. Verify that the set of positive rational numbers does not form a group under division.

4. Show that the set of all three-dimensional vectors forms an infinite Abelian group under vector addition.

5. Prove that all orthogonal matrices of a fixed order form a matrix group. Prove also that all orthogonal matrices of fixed order and of determinant $+1$ form a matrix group.

6. Show that the six functions

$$f_1(x) = x, \qquad f_2(x) = 1-x, \qquad f_3(x) = \frac{x-1}{x},$$

$$f_4(x) = \frac{1}{x}, \qquad f_5(x) = \frac{1}{1-x}, \qquad f_6(x) = \frac{x}{x-1},$$

form a group with the substitution of one function into another as the law of combination.

7. Verify that the six matrices

$$\begin{pmatrix} 1 & 0 \\ 0 & 1 \end{pmatrix}, \quad \begin{pmatrix} -1 & 1 \\ -1 & 0 \end{pmatrix}, \quad \begin{pmatrix} 0 & -1 \\ 1 & -1 \end{pmatrix}, \quad \begin{pmatrix} -1 & 0 \\ 0 & -1 \end{pmatrix},$$
$$\begin{pmatrix} 1 & -1 \\ 1 & 0 \end{pmatrix}, \quad \begin{pmatrix} 0 & 1 \\ -1 & 1 \end{pmatrix}$$

form a matrix group. Show that the group is cyclic and that it may be generated by either of the last two matrices.

8. Show that matrices of the type

$$\begin{pmatrix} a & 0 \\ 0 & 1 \end{pmatrix},$$

where $a \neq 0$, form a matrix group which is isomorphic with the group of real non-zero numbers under multiplication.

9. Show that the four functions

$$f_1(x) = x, \quad f_2(x) = -x, \quad f_3(x) = \frac{1}{x}, \quad f_4(x) = -\frac{1}{x}$$

form a group under substitution of one function into another. Verify that this group is isomorphic with the matrix group whose corresponding elements are

$$\begin{pmatrix} 1 & 0 \\ 0 & 1 \end{pmatrix}, \quad \begin{pmatrix} 1 & 0 \\ 0 & -1 \end{pmatrix}, \quad \begin{pmatrix} -1 & 0 \\ 0 & 1 \end{pmatrix}, \quad \begin{pmatrix} -1 & 0 \\ 0 & -1 \end{pmatrix}.$$

10. Prove that a group of prime order has no proper subgroups and is necessarily cyclic.

11. Examine the structures of groups of order 1 to 4 (inclusive). Show that there are only two possible groups of order 4, one of which is cyclic.

FURTHER READING

Sets

1. C. A. R. BAILEY, *Sets and Logic* 1. Arnold, 1964.
2. J. G. KEMENEY, J. L. SNELL and G. L. THOMPSON, *Introduction to Finite Mathematics*. Prentice-Hall, 1957.
3. L. FELIX, *The Modern Aspect of Mathematics*. Science Editions Inc., New York, 1961.
4. H. G. FLEGG, *Boolean Algebra and its Applications*. Blackie, 1964.
5. F. E. HOHN, *Applied Boolean Algebra: An Elementary Introduction*. Macmillan: New York, 1960.
6. M. PHISTER, Jr., *Logical Design of Digital Computers*. Wiley, 1963 (Chapters 3 and 4).
7. W. FELLER, *Introduction to Probability Theory and Its Applications*. Wiley, 1950 (Chapter 1).
8. J. A. GREEN, *Sets and Groups*. Routledge & Kegan Paul, 1965.

Matrices

1. F. R. GANTMACHER, *Matrix Theory*. Chelsea, 1959 (Vols. 1 and 2).
2. S. PERLIS, *Theory of Matrices*. Addison-Wesley, 1952.
3. R. A. FRAZER, W. J. DUNCAN and A. R. COLLAR, *Elementary Matrices*. Cambridge University Press, 1938.
4. W. L. FERRAR, *Finite Matrices*. Oxford University Press, 1951.
5. G. G. HALL, *Matrices and Tensors*. Pergamon, 1963.
6. R. BELLMAN, *Introduction to Matrix Analysis*. McGraw-Hill, 1960.
7. L. FOX, *An Introduction to Numerical Linear Algebra*. Oxford University Press, 1964.
8. E. BODEWIG, *Matrix Calculus*. North-Holland, 1956.
9. L. PIPES, *Matrix Methods for Engineers*. Prentice-Hall, 1963.
10. H. MARGENAU and G. M. MURPHY, *The Mathematics of Physics and Chemistry*. Van Nostrand, 1956 (Chapter 10).
11. G. STEPHENSON, *Inequalities and Optimal Problems in Mathematics and the Sciences*. Longman, 1971.

Groups

1. W. LEDERMAN, *Introduction to the Theory of Finite Groups*. Oliver and Boyd, 1957.
2. H. MARGENAU and G. M. MURPHY, *The Mathematics of Physics and Chemistry*. Van Nostrand, 1956 (Chapter 15).

3. F. D. MURNAGHAN, *The Theory of Group Representations*. John Hopkins Press, 1938.
4. G. YA. LYABARSKII, *The Application of Group Theory in Physics*. Pergamon, 1960.
5. J. S. LOMONT, *Applications of Finite Groups*. Academic Press, 1959.
6. V. HEINE, *Group Theory*. Pergamon, 1960.
7. H. EYRING, J. WALTER and G. E. KIMBALL, *Quantum Chemistry*. Wiley, 1944 (Chapter 10).
8. F. A. COTTON, *Chemical Applications of Group Theory*. Interscience, 1963.
9. M. TINKHAM, *Group Theory and Quantum Mechanics*. McGraw-Hill, 1964.
10. E. P. WIGNER, *Group Theory and Its Application to the Quantum Mechanics of Atomic Spectra*. Academic Press, 1959.
11. P. ROMAN, *Theory of Elementary Particles*. North-Holland, 1961 (Chapter 1).
12. M. A. JASWON, *An Introduction to Mathematical Crystallography*. Longmans, 1965.
13. J. MATHEWS and R. L. WALKER, *Mathematical Methods of Physics*, Benjamin, 1964 (Chapter 16).

ANSWERS TO PROBLEMS

PROBLEMS 1

1. (*a*) is a finite set.
2. (*a*) is the null set.
3. {1}, {2}, {3}, {4}, {1, 2}, {1, 3}, {1, 4}, {2, 3}, {2, 4}, {3, 4}, {1, 2, 3}, {2, 3, 4}, {1, 2, 4}, {3, 4, 1}, {1, 2, 3, 4} and ø.
4. The sets (*b*), (*c*) and (*e*) are equal.
5. $fg = \begin{pmatrix} 1 & 2 & 3 & 4 \\ 3 & 1 & 4 & 2 \end{pmatrix}, \quad gf = \begin{pmatrix} 1 & 2 & 3 & 4 \\ 4 & 3 & 1 & 2 \end{pmatrix}.$

   There are 4! mappings of which fg and gf are two.
6. The message reads 'you have decoded this message'.
7. $\mathbf{Y} = \begin{pmatrix} y_1 \\ y_2 \\ y_3 \end{pmatrix}, \quad \mathbf{X} = \begin{pmatrix} x_1 \\ x_2 \\ x_3 \end{pmatrix}, \quad \mathbf{A} = \begin{pmatrix} 6 & 2 & -1 \\ 1 & -1 & 2 \\ 7 & 1 & 1 \end{pmatrix}.$

   Since $|\mathbf{A}| = 0$ an inverse transformation does not exist.
8. $\begin{pmatrix} 7 & -1 \\ 6 & -1 \\ -1 & 2 \end{pmatrix}.$

PROBLEMS 2

1. $$\mathbf{A}+\mathbf{B} = \begin{pmatrix} 3 & 3 \\ 7 & 7 \end{pmatrix}, \quad \mathbf{A}-\mathbf{B} = \begin{pmatrix} -1 & 1 \\ -1 & 1 \end{pmatrix},$$

   $$(\mathbf{A}-\mathbf{B})(\mathbf{A}+\mathbf{B}) = \begin{pmatrix} 4 & 4 \\ 4 & 4 \end{pmatrix}, \quad \mathbf{A}^2-\mathbf{B}^2 = \begin{pmatrix} -1 & 5 \\ -5 & 9 \end{pmatrix}$$

2. $$\mathbf{A}+\mathbf{B} = \begin{pmatrix} -1 & 2 & 2 \\ 9 & 7 & 8 \\ 2 & -1 & 3 \end{pmatrix}, \quad \mathbf{A}-\mathbf{B} = \begin{pmatrix} 5 & 0 & 2 \\ -3 & 3 & 6 \\ 0 & 1 & -1 \end{pmatrix},$$

   $$\mathbf{AB} = \begin{pmatrix} 2 & 2 & 5 \\ 28 & 6 & 19 \\ -2 & 0 & 2 \end{pmatrix}, \quad \mathbf{BA} = \begin{pmatrix} -3 & 2 & 1 \\ 19 & 16 & 27 \\ 1 & -4 & -3 \end{pmatrix}.$$

3. $\mathbf{AB} = \begin{pmatrix} 12 & -1 \\ 10 & -9 \end{pmatrix}, \quad \mathbf{BA} = \begin{pmatrix} -2 & 11 \\ 8 & 5 \end{pmatrix}.$

4. $\mathbf{Au} = \begin{pmatrix} 5 \\ 1 \\ -1 \end{pmatrix}$, $\mathbf{A}^2\mathbf{u} = \begin{pmatrix} 12 \\ 2 \\ -7 \end{pmatrix}$, $\mathbf{Av} = \begin{pmatrix} 3 \\ 3 \\ -5 \end{pmatrix}$, $\mathbf{A}^2\mathbf{v} = \begin{pmatrix} 10 \\ 8 \\ -13 \end{pmatrix}$,

$\tilde{\mathbf{u}}\mathbf{A}^2\mathbf{v} = 18$.

8. Symmetric part of **A** is

$$\begin{pmatrix} 1 & 1 & -3 \\ 1 & 0 & \frac{1}{2} \\ -3 & \frac{1}{2} & 2 \end{pmatrix};$$

skew-part of **A** is

$$\begin{pmatrix} 0 & \frac{1}{2} & -2 \\ -\frac{1}{2} & 0 & \frac{1}{4} \\ 2 & -\frac{1}{4} & 0 \end{pmatrix}.$$

11. Skew-Hermitian, symmetric, Hermitian, skew-symmetric, skew-Hermitian.

PROBLEMS 3

3. $\begin{pmatrix} a-ib & -c-id \\ c-id & a+ib \end{pmatrix}$.

5. $\begin{pmatrix} 5 & -3 & \frac{1}{2} \\ -3 & 2 & -\frac{1}{2} \\ 0 & 0 & \frac{1}{2} \end{pmatrix}$.

10. $\begin{pmatrix} 1 & -3 & 2 \\ -3 & 3 & -1 \\ 2 & -1 & 0 \end{pmatrix}$

PROBLEMS 4

1. $x = 3, \quad y = 1, \quad z = 2.$
2. $x = k-1, \quad y = k-1, \quad z = k$, ($k$ arbitrary).
3. $k = 3, \quad -\frac{1}{25}, \quad x = 2, \quad y = 1, \quad z = -4.$
4. (a) $x = 2, \quad y = 4, \quad z = 8.$
   (b) $x = 1, \quad y = 2, \quad z = 3, \quad w = 0.$
   (c) $x = -3k, \quad y = 0, \quad z = k$, ($k$ arbitrary).
   (d) Inconsistent.
   (e) $x = 1+\frac{9}{11}k, \quad y = -\frac{12}{11}k, \quad z = -\frac{8}{11}k, \quad w = k$, ($k$ arbitrary).
   (f) Inconsistent.
   (g) Inconsistent.

6. $x = -151$, $y = 100$; no solution; $x = 153$, $y = -100$; $x = 77$, $y = -50$; $x = 51^2/_3$, $y = -33^1/_3$.
7. $x = \frac{145}{138}$, $y = \frac{286}{138}$.

PROBLEMS 5

1. (a) $\lambda_1 = 3, \quad \mathbf{X}_1 = \begin{pmatrix} -\frac{4}{\sqrt{17}} \\ \frac{1}{\sqrt{17}} \end{pmatrix}; \quad \lambda_2 = 9, \quad \mathbf{X}_2 = \begin{pmatrix} \frac{1}{\sqrt{2}} \\ -\frac{1}{\sqrt{2}} \end{pmatrix}.$

(b) $\lambda_1 = 2, \quad \mathbf{X}_1 = \begin{pmatrix} \frac{1}{\sqrt{2}} \\ -\frac{1}{\sqrt{2}} \end{pmatrix}; \quad \lambda_2 = 3, \quad \mathbf{X}_2 = \begin{pmatrix} 0 \\ 1 \end{pmatrix}.$

(c) $\lambda_1 = 1, \quad \mathbf{X}_1 = \begin{pmatrix} 1 \\ 0 \\ 0 \end{pmatrix}; \quad \lambda_2 = 2, \quad \mathbf{X}_2 = \begin{pmatrix} \frac{4}{\sqrt{17}} \\ \frac{1}{\sqrt{17}} \\ 0 \end{pmatrix};$

$\lambda_3 = 3, \quad \mathbf{X}_3 = \begin{pmatrix} \frac{29}{\sqrt{989}} \\ \frac{12}{\sqrt{989}} \\ \frac{2}{\sqrt{989}} \end{pmatrix}.$

(d) $\lambda_1 = 2, \quad \mathbf{X}_1 = \begin{pmatrix} 1 \\ 0 \\ 0 \end{pmatrix}; \quad \lambda_2 = \lambda_3 = 1, \quad \mathbf{X}_2 = \mathbf{X}_3 = \begin{pmatrix} -\frac{3}{\sqrt{10}} \\ \frac{1}{\sqrt{10}} \\ 0 \end{pmatrix}.$

2. (a) $\lambda_1 = -1, \quad \mathbf{X}_1 = \begin{pmatrix} \frac{1}{\sqrt{2}} \\ -\frac{1}{\sqrt{2}} \\ 0 \end{pmatrix}; \quad \lambda_2 = 1, \quad \mathbf{X}_2 = \begin{pmatrix} \frac{1}{\sqrt{2}} \\ \frac{1}{\sqrt{2}} \\ 0 \end{pmatrix};$

$$\lambda_3 = 1, \quad \mathbf{X}_3 = \begin{pmatrix} 0 \\ 0 \\ 1 \end{pmatrix}.$$

(b) $$\lambda_1 = 0, \quad \mathbf{X}_1 = \begin{pmatrix} \frac{1}{\sqrt{2}} \\ -\frac{1}{\sqrt{2}} \\ 0 \end{pmatrix}; \quad \lambda_2 = 4, \quad \mathbf{X}_2 = \begin{pmatrix} \frac{1}{\sqrt{2}} \\ \frac{1}{\sqrt{2}} \\ 0 \end{pmatrix};$$

$$\lambda_3 = 1, \quad \mathbf{X}_3 = \begin{pmatrix} 0 \\ 0 \\ 1 \end{pmatrix}.$$

(c) $$\lambda_1 = 0, \quad \mathbf{X}_1 = \begin{pmatrix} -\frac{2}{3} \\ \frac{2}{3} \\ \frac{1}{3} \end{pmatrix}; \quad \lambda_2 = 6, \quad \mathbf{X}_2 = \begin{pmatrix} \frac{2}{3} \\ \frac{1}{3} \\ \frac{2}{3} \end{pmatrix};$$

$$\lambda_3 = 3, \quad \mathbf{X}_3 = \begin{pmatrix} \frac{1}{3} \\ \frac{2}{3} \\ -\frac{2}{3} \end{pmatrix}.$$

(d) $$\lambda_1 = 0, \quad \mathbf{X}_1 = \begin{pmatrix} -(1+i)/\sqrt{3} \\ \frac{1}{\sqrt{3}} \end{pmatrix};$$

$$\lambda_2 = 3, \quad \mathbf{X}_2 = \begin{pmatrix} (1+i)/\sqrt{6} \\ \sqrt{\frac{2}{3}} \end{pmatrix}.$$

9. $\lambda = \pm 1, \quad \pm i.$

PROBLEMS 6

1. (a) $\begin{pmatrix} 5 & 0 \\ 0 & -2 \end{pmatrix}$, (*b*) $\begin{pmatrix} 1 & 0 & 0 \\ 0 & 2 & 0 \\ 0 & 0 & 3 \end{pmatrix}$,

(c) $\begin{pmatrix} 0 & 0 & 0 \\ 0 & 3(1+i) & 0 \\ 0 & 0 & 3(1-i) \end{pmatrix}$.

2. (a) $\begin{pmatrix} 5 & 0 \\ 0 & -5 \end{pmatrix}$, (b) $\begin{pmatrix} 1 & 0 & 0 \\ 0 & 1 & 0 \\ 0 & 0 & -1 \end{pmatrix}$,

(c) $\begin{pmatrix} 0 & 0 & 0 \\ 0 & 1 & 0 \\ 0 & 0 & 4 \end{pmatrix}$, (d) $\begin{pmatrix} 1 & 0 & 0 & 0 \\ 0 & 1 & 0 & 0 \\ 0 & 0 & -1 & 0 \\ 0 & 0 & 0 & -1 \end{pmatrix}$.

3. (a) $\begin{pmatrix} 4 & 0 \\ 0 & -5 \end{pmatrix}$, (b) $\begin{pmatrix} 5 & 0 \\ 0 & 2 \end{pmatrix}$, (c) $\begin{pmatrix} 1 & 0 & 0 \\ 0 & \frac{5+\sqrt{33}}{2} & 0 \\ 0 & 0 & \frac{5-\sqrt{33}}{2} \end{pmatrix}$

4. (a) $\begin{pmatrix} 2 & -\frac{5}{2} \\ -\frac{5}{2} & 5 \end{pmatrix}$, (b) $\begin{pmatrix} 1 & -1 & 0 \\ -1 & 2 & -1 \\ 0 & -1 & 2 \end{pmatrix}$, (c) $\begin{pmatrix} 1 & 4 & -5 \\ 4 & 0 & 0 \\ -5 & 0 & 2 \end{pmatrix}$.

5. (a) $2u_1^2+\frac{15}{8}u_2^2, \quad u_1 = x_1-\frac{5}{4}x_2, \quad u_2 = x_2.$

(b) $u_1^2+u_2^2+u_3^2, \quad u_1 = x_1-x_2, \quad u_2 = x_2-x_3, \quad u_3 = x_3.$

(c) $u_1^2+2u_2^2-u_3^2, \quad u_1 = x_1+4x_2-5x_3, \quad u_2 = x_3,$
$u_3 = 4x_2-5x_3.$

. Exact eigenvalues are 2, $\frac{5\pm\sqrt{5}}{2}$.

PROBLEMS 7

1. $\frac{1}{2}\begin{pmatrix} \alpha & \beta \\ \beta & \alpha \end{pmatrix}$, where $\alpha = 4^{14}+2^{14}$, $\beta = 4^{14}-2^{14}$; $\begin{pmatrix} 1 & 0 \\ 90 & 1 \end{pmatrix}$.

4. $\begin{pmatrix} -12099 & 12355 & 6305 \\ -12100 & 12356 & 6305 \\ -13120 & 13120 & 6561 \end{pmatrix}$.

12. $x_r = 2^r$.

INDEX

Index

A CATALOG OF SELECTED

DOVER BOOKS

IN ALL FIELDS OF INTEREST

A CATALOG OF SELECTED DOVER BOOKS IN ALL FIELDS OF INTEREST

DRAWINGS OF REMBRANDT, edited by Seymour Slive. Updated Lippmann, Hofstede de Groot edition, with definitive scholarly apparatus. All portraits, biblical sketches, landscapes, nudes. Oriental figures, classical studies, together with selection of work by followers. 550 illustrations. Total of 630pp. 9⅛ × 12¼.
21485-0, 21486-9 Pa., Two-vol. set $25.00

GHOST AND HORROR STORIES OF AMBROSE BIERCE, Ambrose Bierce. 24 tales vividly imagined, strangely prophetic, and decades ahead of their time in technical skill: "The Damned Thing," "An Inhabitant of Carcosa," "The Eyes of the Panther," "Moxon's Master," and 20 more. 199pp. 5⅜ × 8½. 20767-6 Pa. $3.95

ETHICAL WRITINGS OF MAIMONIDES, Maimonides. Most significant ethical works of great medieval sage, newly translated for utmost precision, readability. Laws Concerning Character Traits, Eight Chapters, more. 192pp. 5⅜ × 8½.
24522-5 Pa. $4.50

THE EXPLORATION OF THE COLORADO RIVER AND ITS CANYONS, J. W. Powell. Full text of Powell's 1,000-mile expedition down the fabled Colorado in 1869. Superb account of terrain, geology, vegetation, Indians, famine, mutiny, treacherous rapids, mighty canyons, during exploration of last unknown part of continental U.S. 400pp. 5⅜ × 8½. 20094-9 Pa. $6.95

HISTORY OF PHILOSOPHY, Julián Marías. Clearest one-volume history on the market. Every major philosopher and dozens of others, to Existentialism and later. 505pp. 5⅜ × 8½. 21739-6 Pa. $8.50

ALL ABOUT LIGHTNING, Martin A. Uman. Highly readable non-technical survey of nature and causes of lightning, thunderstorms, ball lightning, St. Elmo's Fire, much more. Illustrated. 192pp. 5⅜ × 8½. 25237-X Pa. $5.95

SAILING ALONE AROUND THE WORLD, Captain Joshua Slocum. First man to sail around the world, alone, in small boat. One of great feats of seamanship told in delightful manner. 67 illustrations. 294pp. 5⅜ × 8½. 20326-3 Pa. $4.95

LETTERS AND NOTES ON THE MANNERS, CUSTOMS AND CONDITIONS OF THE NORTH AMERICAN INDIANS, George Catlin. Classic account of life among Plains Indians: ceremonies, hunt, warfare, etc. 312 plates. 572pp. of text. 6⅛ × 9¼. 22118-0, 22119-9 Pa. Two-vol. set $15.90

ALASKA: The Harriman Expedition, 1899, John Burroughs, John Muir, et al. Informative, engrossing accounts of two-month, 9,000-mile expedition. Native peoples, wildlife, forests, geography, salmon industry, glaciers, more. Profusely illustrated. 240 black-and-white line drawings. 124 black-and-white photographs. 3 maps. Index. 576pp. 5⅜ × 8½. 25109-8 Pa. $11.95

HOW TO WRITE, Gertrude Stein. Gertrude Stein claimed anyone could understand her unconventional writing—here are clues to help. Fascinating improvisations, language experiments, explanations illuminate Stein's craft and the art of writing. Total of 414pp. 4⅜ × 6⅜. 23144-5 Pa. $5.95

ADVENTURES AT SEA IN THE GREAT AGE OF SAIL: Five Firsthand Narratives, edited by Elliot Snow. Rare true accounts of exploration, whaling, shipwreck, fierce natives, trade, shipboard life, more. 33 illustrations. Introduction. 353pp. 5⅜ × 8½. 25177-2 Pa. $7.95

THE HERBAL OR GENERAL HISTORY OF PLANTS, John Gerard. Classic descriptions of about 2,850 plants—with over 2,700 illustrations—includes Latin and English names, physical descriptions, varieties, time and place of growth, more. 2,706 illustrations. xlv + 1,678pp. 8½ × 12¼. 23147-X Cloth. $75.00

DOROTHY AND THE WIZARD IN OZ, L. Frank Baum. Dorothy and the Wizard visit the center of the Earth, where people are vegetables, glass houses grow and Oz characters reappear. Classic sequel to *Wizard of Oz*. 256pp. 5⅜ × 8. 24714-7 Pa. $4.95

SONGS OF EXPERIENCE: Facsimile Reproduction with 26 Plates in Full Color, William Blake. This facsimile of Blake's original "Illuminated Book" reproduces 26 full-color plates from a rare 1826 edition. Includes "The Tyger," "London," "Holy Thursday," and other immortal poems. 26 color plates. Printed text of poems. 48pp. 5¼ × 7. 24636-1 Pa. $3.50

SONGS OF INNOCENCE, William Blake. The first and most popular of Blake's famous "Illuminated Books," in a facsimile edition reproducing all 31 brightly colored plates. Additional printed text of each poem. 64pp. 5¼ × 7. 22764-2 Pa. $3.50

PRECIOUS STONES, Max Bauer. Classic, thorough study of diamonds, rubies, emeralds, garnets, etc.: physical character, occurrence, properties, use, similar topics. 20 plates, 8 in color. 94 figures. 659pp. 6⅛ × 9¼. 21910-0, 21911-9 Pa., Two-vol. set $15.90

ENCYCLOPEDIA OF VICTORIAN NEEDLEWORK, S. F. A. Caulfeild and Blanche Saward. Full, precise descriptions of stitches, techniques for dozens of needlecrafts—most exhaustive reference of its kind. Over 800 figures. Total of 679pp. 8⅛ × 11. Two volumes. Vol. 1 22800-2 Pa. $11.95
Vol. 2 22801-0 Pa. $11.95

THE MARVELOUS LAND OF OZ, L. Frank Baum. Second Oz book, the Scarecrow and Tin Woodman are back with hero named Tip, Oz magic. 136 illustrations. 287pp. 5⅜ × 8½. 20692-0 Pa. $5.95

WILD FOWL DECOYS, Joel Barber. Basic book on the subject, by foremost authority and collector. Reveals history of decoy making and rigging, place in American culture, different kinds of decoys, how to make them, and how to use them. 140 plates. 156pp. 7⅞ × 10¾. 20011-6 Pa. $8.95

HISTORY OF LACE, Mrs. Bury Palliser. Definitive, profusely illustrated chronicle of lace from earliest times to late 19th century. Laces of Italy, Greece, England, France, Belgium, etc. Landmark of needlework scholarship. 266 illustrations. 672pp. 6⅛ × 9¼. 24742-2 Pa. $14.95

PLANTS OF THE BIBLE, Harold N. Moldenke and Alma L. Moldenke. Standard reference to all 230 plants mentioned in Scriptures. Latin name, biblical reference, uses, modern identity, much more. Unsurpassed encyclopedic resource for scholars, botanists, nature lovers, students of Bible. Bibliography. Indexes. 123 black-and-white illustrations. 384pp. 6 × 9. 25069-5 Pa. $8.95

FAMOUS AMERICAN WOMEN: A Biographical Dictionary from Colonial Times to the Present, Robert McHenry, ed. From Pocahontas to Rosa Parks, 1,035 distinguished American women documented in separate biographical entries. Accurate, up-to-date data, numerous categories, spans 400 years. Indices. 493pp. 6½ × 9¼. 24523-3 Pa. $9.95

THE FABULOUS INTERIORS OF THE GREAT OCEAN LINERS IN HISTORIC PHOTOGRAPHS, William H. Miller, Jr. Some 200 superb photographs capture exquisite interiors of world's great "floating palaces"—1890's to 1980's: *Titanic, Ile de France, Queen Elizabeth, United States, Europa,* more. Approx. 200 black-and-white photographs. Captions. Text. Introduction. 160pp. 8⅝ × 11¾. 24756-2 Pa. $9.95

THE GREAT LUXURY LINERS, 1927–1954: A Photographic Record, William H. Miller, Jr. Nostalgic tribute to heyday of ocean liners. 186 photos of Ile de France, Normandie, Leviathan, Queen Elizabeth, United States, many others. Interior and exterior views. Introduction. Captions. 160pp. 9 × 12. 24056-8 Pa. $9.95

A NATURAL HISTORY OF THE DUCKS, John Charles Phillips. Great landmark of ornithology offers complete detailed coverage of nearly 200 species and subspecies of ducks: gadwall, sheldrake, merganser, pintail, many more. 74 full-color plates, 102 black-and-white. Bibliography. Total of 1,920pp. 8⅝ × 11¼. 25141-1, 25142-X Cloth. Two-vol. set $100.00

THE SEAWEED HANDBOOK: An Illustrated Guide to Seaweeds from North Carolina to Canada, Thomas F. Lee. Concise reference covers 78 species. Scientific and common names, habitat, distribution, more. Finding keys for easy identification. 224pp. 5⅜ × 8½. 25215-9 Pa. $5.95

THE TEN BOOKS OF ARCHITECTURE: The 1755 Leoni Edition, Leon Battista Alberti. Rare classic helped introduce the glories of ancient architecture to the Renaissance. 68 black-and-white plates. 336pp. 8⅜ × 11¼. 25239-6 Pa. $14.95

MISS MACKENZIE, Anthony Trollope. Minor masterpieces by Victorian master unmasks many truths about life in 19th-century England. First inexpensive edition in years. 392pp. 5⅜ × 8½. 25201-9 Pa. $7.95

THE RIME OF THE ANCIENT MARINER, Gustave Doré, Samuel Taylor Coleridge. Dramatic engravings considered by many to be his greatest work. The terrifying space of the open sea, the storms and whirlpools of an unknown ocean, the ice of Antarctica, more—all rendered in a powerful, chilling manner. Full text. 38 plates. 77pp. 9¼ × 12. 22305-1 Pa. $4.95

THE EXPEDITIONS OF ZEBULON MONTGOMERY PIKE, Zebulon Montgomery Pike. Fascinating first-hand accounts (1805–6) of exploration of Mississippi River, Indian wars, capture by Spanish dragoons, much more. 1,088pp. 5⅜ × 8½. 25254-X, 25255-8 Pa. Two-vol. set $23.90

DEGAS: An Intimate Portrait, Ambroise Vollard. Charming, anecdotal memoir by famous art dealer of one of the greatest 19th-century French painters. 14 black-and-white illustrations. Introduction by Harold L. Van Doren. 96pp. 5⅜ × 8½.
25131-4 Pa. $3.95

PERSONAL NARRATIVE OF A PILGRIMAGE TO ALMANDINAH AND MECCAH, Richard Burton. Great travel classic by remarkably colorful personality. Burton, disguised as a Moroccan, visited sacred shrines of Islam, narrowly escaping death. 47 illustrations. 959pp. 5⅜ × 8½. 21217-3, 21218-1 Pa., Two-vol. set $17.90

PHRASE AND WORD ORIGINS, A. H. Holt. Entertaining, reliable, modern study of more than 1,200 colorful words, phrases, origins and histories. Much unexpected information. 254pp. 5⅜ × 8½. 20758-7 Pa. $5.95

THE RED THUMB MARK, R. Austin Freeman. In this first Dr. Thorndyke case, the great scientific detective draws fascinating conclusions from the nature of a single fingerprint. Exciting story, authentic science. 320pp. 5⅜ × 8½. (Available in U.S. only) 25210-8 Pa. $5.95

AN EGYPTIAN HIEROGLYPHIC DICTIONARY, E. A. Wallis Budge. Monumental work containing about 25,000 words or terms that occur in texts ranging from 3000 B.C. to 600 A.D. Each entry consists of a transliteration of the word, the word in hieroglyphs, and the meaning in English. 1,314pp. 6⅛ × 10.
23615-3, 23616-1 Pa., Two-vol. set $27.90

THE COMPLEAT STRATEGYST: Being a Primer on the Theory of Games of Strategy, J. D. Williams. Highly entertaining classic describes, with many illustrated examples, how to select best strategies in conflict situations. Prefaces. Appendices. xvi + 268pp. 5⅜ × 8½. 25101-2 Pa. $5.95

THE ROAD TO OZ, L. Frank Baum. Dorothy meets the Shaggy Man, little Button-Bright and the Rainbow's beautiful daughter in this delightful trip to the magical Land of Oz. 272pp. 5⅜ × 8. 25208-6 Pa. $4.95

POINT AND LINE TO PLANE, Wassily Kandinsky. Seminal exposition of role of point, line, other elements in non-objective painting. Essential to understanding 20th-century art. 127 illustrations. 192pp. 6½ × 9¼. 23808-3 Pa. $4.50

LADY ANNA, Anthony Trollope. Moving chronicle of Countess Lovel's bitter struggle to win for herself and daughter Anna their rightful rank and fortune—perhaps at cost of sanity itself. 384pp. 5⅜ × 8½. 24669-8 Pa. $6.95

EGYPTIAN MAGIC, E. A. Wallis Budge. Sums up all that is known about magic in Ancient Egypt: the role of magic in controlling the gods, powerful amulets that warded off evil spirits, scarabs of immortality, use of wax images, formulas and spells, the secret name, much more. 253pp. 5⅜ × 8½. 22681-6 Pa. $4.50

THE DANCE OF SIVA, Ananda Coomaraswamy. Preeminent authority unfolds the vast metaphysic of India: the revelation of her art, conception of the universe, social organization, etc. 27 reproductions of art masterpieces. 192pp. 5⅜ × 8½.
24817-8 Pa. $5.95

CATALOG OF DOVER BOOKS

AMERICAN CLIPPER SHIPS: 1833-1858, Octavius T. Howe & Frederick C. Matthews. [illegible] encyclopedic review of [illegible] clipper ships from [illegible] Index. [illegible] 5⅜ × 8½. [illegible] Two-volume set [illegible]

TOWARDS A NEW ARCHITECTURE, Le Corbusier. [illegible] legendary founder [illegible] aesthetic [illegible] views on [illegible] relation of form to function [illegible] mass-production [illegible] illustrated. Unabridged translation of the [illegible] edition. Introduction by Frederick Etchells. [illegible] 6⅛ × 9¼. [illegible]

THE BOOK OF KELLS, edited by Blanche Cirker. [illegible] full-page plates [illegible] the greatest illuminated manuscript of the Middle Ages, [illegible] Publisher's Note. Captions. 32pp. 9⅜ × 12¼. [illegible]

[illegible] FROM [illegible] plus 17 [illegible] Canyon [illegible] The Strange [illegible] 208pp. 5⅜ × 8½. [illegible]

[illegible] Charles [illegible] Davis [illegible] 0-486-[illegible]

[illegible] MATHEMATICAL PUZZLES, [illegible] author [illegible] money, speed, [illegible] complete solutions. [illegible]

THE WILL TO BELIEVE, HUMAN IMMORTALITY, William James. Two books bound together. [illegible] arguments for human immortality. [illegible] 5⅜ × 8½. [illegible]

THE [illegible] MARE, [illegible] R. van Gulik. [illegible] 5⅜ × 8½. [illegible]

[illegible] JUDGE DEE [illegible] [illegible]

[illegible] [illegible] catalog. [illegible]

[illegible] books each year on science, elementary and advanced mathematics, biology, music, art, literary history, social sciences and other areas.